T0305802

Mathematical Aspects of Quantum Computing 2007

Kinki University Series on Quantum Computing

Editor-in-Chief: Mikio Nakahara *(Kinki University, Japan)*

ISSN: 1793-7299

Published

Vol. 1   Mathematical Aspects of Quantum Computing 2007
*edited by Mikio Nakahara, Robabeh Rahimi (Kinki Univ., Japan) &
Akira SaiToh (Osaka Univ., Japan)*

Kinki University Series on Quantum Computing – Vol. 1

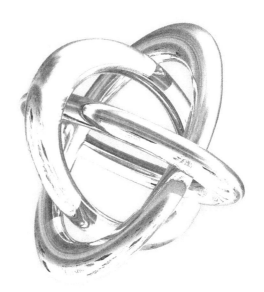

editors

Mikio Nakahara
Robabeh Rahimi
Kinki University, Japan

Akira SaiToh
Osaka University, Japan

Mathematical Aspects of Quantum Computing 2007

World Scientific

NEW JERSEY · LONDON · SINGAPORE · BEIJING · SHANGHAI · HONG KONG · TAIPEI · CHENNAI

Published by

World Scientific Publishing Co. Pte. Ltd.

5 Toh Tuck Link, Singapore 596224

USA office: 27 Warren Street, Suite 401-402, Hackensack, NJ 07601

UK office: 57 Shelton Street, Covent Garden, London WC2H 9HE

British Library Cataloguing-in-Publication Data
A catalogue record for this book is available from the British Library.

MATHEMATICAL ASPECTS OF QUANTUM COMPUTING 2007
Kinki University Series on Quantum Computing — Vol. 1

Copyright © 2008 by World Scientific Publishing Co. Pte. Ltd.

ISBN-13 978-981-281-447-0
ISBN-10 981-281-447-7

Printed in Singapore.

PREFACE

This volume contains lecture notes and poster contributions presented at the summer school "Mathematical Aspects of Quantum Computing", held from 27 to 29 August, 2007 at Kinki University in Osaka, Japan. The aims of this summer school were to exchange and share ideas among researchers working in various fields of quantum computing particularly from a mathematical point of view, and to motivate students to tackle the advanced subjects of quantum computing.

Invited speakers to the summer school were asked to prepare their lecture notes in a self-contained way and therefore we expect that each contribution will be useful for students and researchers even with less background to the topic.

This summer school was supported by "Open Research Center" Project for Private Universities: matching fund subsidy from MEXT (Ministry of Education, Culture, Sports, Science and Technology).

We yearn for continued outstanding success in this field and wish to expand summer schools and conferences further in the field of quantum computing based at Kinki University.

Finally, we would like to thank Ms Zhang Ji of World Scientific for her excellent editorial work.

Osaka, January 2008

Mikio Nakahara
Robabeh Rahimi
Akira SaiToh

Summer School on
Mathematical Aspects of Quantum Computing

Kinki Univerisity, Osaka, Japan

---

27 – 29   August   2007

---

27 August

Mikio Nakahara (Kinki Univerisity, Japan)
Quantum Computing: An Overview

Kazuhiro Sakuma (Kinki University, Japan)
Braid Group and Topological Quantum Computing

28 August

Damian J. H. Markham (Tokyo University, Japan)
An Introduction to Entanglement Theory

Shogo Tanimura (Kyoto University, Japan)
Holonomic Quantum Computing and Its Optimization

29 August

Sahin Kaya Ozdemir (Osaka University, Japan)
Playing Games in Quantum Mechanical Settings: Features of Quantum Games

Manabu Hagiwara (National Institute of Advanced Industrial Science and Technology, Japan)
Quantum Error Correction Codes

Poster Presentations

Vahideh Ebrahimi (Kinki University, Japan)
Controlled Teleportation of an Arbitrary Unknown Two-Qubit Entangled State

Yukihiro Ota (Kinki University, Japan)
Notes on Dür-Cirac Classification

Robabeh Rahimi (Kinki University, Japan)
Bang-Bang Control of Entanglement in Spin-Bus-Boson Model

Akira SaiToh (Osaka University, Japan)
Numerical Computation of Time-Dependent Multipartite Nonclassical Correlation

Takuya Yamano (Ochanomizu University, Japan)
On Classical No-Cloning Theorem under Liouville Dynamics and Distances

LIST OF PARTICIPANTS

Aoki, Takashi	Kinki University, Japan
Chinen, Koji	Osaka City University, Japan
Ebrahimi Bakhtavar, Vahideh	Kinki University, Japan
Hagiwara, Manabu	National Institute of Advanced Industrial Science and Technology, Japan
Ikeda, Yasushi	Kyoto University, Japan
Inoue, Kaiki Taro	Kinki University, Japan
Kikuta, Toshiyuki	Kinki University, Japan
Kobata, Kumi	Kinki University, Japan
Kondo, Yasushi	Kinki University, Japan
Markham, Damian James Harold	Tokyo University, Japan
Minami, Kaori	Kinki University, Japan
Nagaoka, Shoyu	Kinki University, Japan
Nakagawa, Nobuo	Kinki University, Japan
Nakahara, Mikio	Kinki University, Japan
Ootsuka, Takayoshi	Kinki University, Japan
Ota, Yukihiro	Kinki University, Japan
Ozdemir, Sahin Kaya	Osaka University, Japan
Rahimi Darabad, Robabeh	Kinki University, Japan
SaiToh, Akira	Osaka University, Japan
Sakuma, Kazuhiro	Kinki University, Japan
Sasano, Hiroshi	Kinki University, Japan
Segawa, Etsuo	Yokohama National University, Japan
Sugita, Ayumu	Osaka City University, Japan
Tanimura, Shogo	Kyoto University, Japan
Tomita, Hiroyuki	Kinki University, Japan
Unoki, Makoto	Waseda University, Japan
Yamano, Takuya	Ochanomizu University, Japan
Yamanouchi, Akiko	ITOCHU Techno-Solutions Corporations, Japan
Yoshida, Motoyuki	Waseda University, Japan

CONTENTS

Preface   v

Quantum Computing: An Overview   1
   M. Nakahara

Braid Group and Topological Quantum Computing   55
   T. Ootsuka, K. Sakuma

An Introduction to Entanglement Theory   91
   D. J. H. Markham

Holonomic Quantum Computing and Its Optimization   115
   S. Tanimura

Playing Games in Quantum Mechanical Settings: Features of
Quantum Games   139
   Ş. K. Özdemir, J. Shimamura, N. Imoto

Quantum Error-Correcting Codes   181
   M. Hagiwara

—**Poster Summaries**—
Controled Teleportation of an Arbitrary Unknown Two-Qubit
Entangled State   213
   V. Ebrahimi, R. Rahimi, M. Nakahara

Notes on the Dür–Cirac Classification   215
   Y. Ota, M. Yoshida, I. Ohba

Bang-Bang Control of Entanglement in Spin-Bus-Boson Model   217
  R. Rahimi, A. SaiToh, M. Nakahara

Numerical Computation of Time-Dependent Multipartite
Nonclassical Correlation   219
  A. SaiToh, R. Rahimi, M. Nakahara, M. Kitagawa

On Classical No-Cloning Theorem Under Liouville Dynamics
and Distances   221
  T. Yamano, O. Iguchi

QUANTUM COMPUTING: AN OVERVIEW

MIKIO NAKAHARA

Department of Physics, Kinki University, Higashi-Osaka 577-8502, Japan
E-mail: nakahara@math.kindai.ac.jp

Elements of quantum computing and quantum infromation processing are introduced for nonspecialists. Subjects inclulde quantum physics, qubits, quantum gates, quantum algorithms, decoherece, quantum error correcting codes and physical realizations. Presentations of these subjects are as pedagogical as possible. Some sections are meant to be brief introductions to contributions by other lecturers.

Keywords: Quantum Physics, Qubits, Quantum Gates, Quantum Algorithms.

1. Introduction

Quantum computing and quantum information processing are emerging disciplines in which the principles of quantum physics are employed to store and process information. We use the classical digital technology at almost every moment in our lives: computers, mobile phones, mp3 players, just to name a few. Even though quantum mechanics is used in the design of devices such as LSI, the logic is purely classical. This means that an AND circuit, for example, produces *definitely* 1 when the inputs are 1 and 1. One of the most remarkable aspects of the principles of quantum physics is the *superposition principle* by which a quantum system can take several different states *simultaneously*. The input for a quantum computing device may be a superposition of many possible inputs, and accordingly the output is also a superposition of the corresponding output states. Another aspect of quantum physics, which is far beyond the classical description, is *entanglement*. Given several objects in a classical world, they can be described by specifying each object separately. Given a group of five people, for example, this group can be described by specifying the height, color of eyes, personality and so on of each constituent person. In a quantum world, however, only a very small subset of all possible states can be described by such individual specifications. In other words, most quantum states cannot be described by

2

such individual specifications, thereby being called "entangled". Why and how these two features give rise to the enormous computational power in quantum computing and quantum information processing will be explained in this contribution.

A part of this lecture note is based on our forthcoming book.[1] General references are [2–4].

2. Quantum Physics

2.1. *Notation and conventions*

We will exclusively work with a finite-dimensional complex vector space $\mathbb{C}^n$ with an inner product $\langle \ , \ \rangle$ (Hilbert spaces). A vector in $\mathbb{C}^n$ is called a ket vector or a ket and is denoted as

$$|x\rangle = \begin{pmatrix} x_1 \\ \vdots \\ x_n \end{pmatrix} \quad x_i \in \mathbb{C}$$

while a vector in the dual space $\mathbb{C}^{n*}$ is called a bra vector or a bra and denoted

$$\langle \alpha | = (\alpha_1, \ldots, \alpha_n) \quad \alpha_i \in \mathbb{C}.$$

Index i sometimes runs from 0 to $n-1$. The inner product of $|x\rangle$ and $\langle \alpha |$ is

$$\langle \alpha | x \rangle = \sum_{i=1}^{n} \alpha_i x_i.$$

This inner product naturally introduces a correspondence

$$|x\rangle = (x_1, \ldots, x_n)^t \leftrightarrow \langle x | = (x_1^*, \ldots, x_n^*),$$

by which an inner product of two vectors are defined as $\langle x | y \rangle = \sum_{i=1}^{n} x_i^* y_i$. The inner product naturally defines the norm of a vector $|x\rangle$ as $\||x\rangle\| = \sqrt{\langle x | x \rangle}$.

Pauli matrices are generators of $\mathfrak{su}(2)$ and denoted

$$\sigma_x = \begin{pmatrix} 0 & 1 \\ 1 & 0 \end{pmatrix}, \ \sigma_y = \begin{pmatrix} 0 & -i \\ i & 0 \end{pmatrix}, \ \sigma_z = \begin{pmatrix} 1 & 0 \\ 0 & -1 \end{pmatrix}$$

in the basis in which σ_z is diagonalized. Symbols $X = \sigma_x, Y = -i\sigma_y$ and $Z = \sigma_z$ are also employed.

Let A be an $m \times n$ matrix and B be a $p \times q$ matrix. Then

$$A \otimes B = \begin{pmatrix} a_{11}B, a_{12}B, \ldots, a_{1n}B \\ a_{21}B, a_{22}B, \ldots, a_{2n}B \\ \cdots \\ a_{m1}B, a_{m2}B, \ldots, a_{mn}B \end{pmatrix}$$

is an $(mp) \times (nq)$ matrix called the tensor product of A and B.

2.2. Axioms of quantum mechanics

Quantum mechanics was discovered roughly a century ago.[5–10] In spite of its long history, the interpretation of the wave function remains an open question. Here we adopt the most popular one, called the Copenhagen interpretation.

A 1 A pure state in quantum mechanics is represented by a normalized vector $|\psi\rangle$ in a Hilbert space $\mathcal{H}$ associated with the system. If two states $|\psi_1\rangle$ and $|\psi_2\rangle$ are physical states of the system, their linear superposition $c_1|\psi_1\rangle + c_2|\psi_2\rangle$ ($c_k \in \mathbb{C}$), with $\sum_{i=1}^{2} |c_i|^2 = 1$, is also a possible state of the same system (superposition principle).

A 2 For any physical quantity (observable) a, there exists a corresponding Hermitian operator A acting on $\mathcal{H}$. When a measurement of a is made, the outcome is one of the eigenvalues λ_j of A. Let λ_1 and λ_2 be two eigenvalues of A: $A|\lambda_i\rangle = \lambda_i|\lambda_i\rangle$. Consider a superposition state $c_1|\lambda_1\rangle + c_2|\lambda_2\rangle$. If we measure a in this state, the state undergoes an abrupt change (wave function collapse) to one of the eigenstates $|\lambda_i\rangle$ corresponding to the observed eigenvalue λ_i. Suppose we prepare many copies of the state $c_1|\lambda_1\rangle + c_2|\lambda_2\rangle$. The probability of collapsing to the state $|\lambda_i\rangle$ is given by $|c_i|^2$ ($i = 1, 2$). The complex coefficient c_i is called the probability amplitude in this sense. It should be noted that a measurement produces one outcome λ_i and the probability of obtaining it is experimentally evaluated only after repeating measurements with many copies of the same state. These statements are easily generalized to superposition states of more than two states.

A 3 The time dependence of a state is governed by the Schrödinger equation

$$i\hbar \frac{\partial |\psi\rangle}{\partial t} = H|\psi\rangle, \tag{1}$$

where $\hbar$ is a physical constant known as the Planck constant and H is a Hermitian operator (matrix) corresponding to the energy of the system and is called the Hamiltonian.

Several comments are in order.

- In Axiom A 1, the phase of the vector may be chosen arbitrarily; $|\psi\rangle$ in fact represents the "ray" $\{e^{i\alpha}|\psi\rangle \ |\alpha \in \mathbb{R}\}$. This is called the ray representation. The overall phase is not observable and has no physical meaning.

- Axiom A 2 may be formulated in a different but equivalent way as follows. Suppose we would like to measure an observable a. Let the spectral decomposition of the corresponding operator A be

$$A = \sum_i \lambda_i |\lambda_i\rangle\langle\lambda_i|, \text{ where } A|\lambda_i\rangle = \lambda_i|\lambda_i\rangle.$$

Then the expectation value $\langle A\rangle$ of a after measurements with respect to many copies of $|\psi\rangle$ is

$$\langle A\rangle = \langle\psi|A|\psi\rangle. \tag{2}$$

Let us expand $|\psi\rangle$ in terms of $|\lambda_i\rangle$ as $|\psi\rangle = \sum_i c_i|\lambda_i\rangle$. According to A 2, the probability of observing λ_i upon measurement of a is $|c_i|^2$ and therefore the expectation value after many measurements is $\sum_i \lambda_i|c_i|^2$. If, conversely, Eq. (2) is employed, we will obtain the same result since

$$\langle\psi|A|\psi\rangle = \sum_{i,j} c_j^* c_i \langle\lambda_j|A|\lambda_i\rangle = \sum_{i,j} \lambda_i c_j^* c_i \delta_{ij} = \sum_i \lambda_i|c_i|^2.$$

This measurement is called the projective measurement. Any particular outcome λ_i will be found with the probability

$$|c_i|^2 = \langle\psi|P_i|\psi\rangle, \tag{3}$$

where $P_i = |\lambda_i\rangle\langle\lambda_i|$ is the projection operator and the state immediately after the measurement is $|\lambda_i\rangle$ or equivalently

$$P_i|\psi\rangle/\sqrt{\langle\psi|P_i|\psi\rangle}. \tag{4}$$

- The Schrödinger equation (1) in Axiom A 3 is formally solved to yield

$$|\psi(t)\rangle = e^{-iHt/\hbar}|\psi(0)\rangle, \tag{5}$$

if the Hamiltonian H is time-independent, while

$$|\psi(t)\rangle = \mathcal{T}\exp\left[-\frac{i}{\hbar}\int_0^t H(t)dt\right]|\psi(0)\rangle \tag{6}$$

if H depends on t, where $\mathcal{T}$ is the time-ordering operator. The state at $t > 0$ is $|\psi(t)\rangle = U(t)|\psi(0)\rangle$. The operator $U(t) : |\psi(0)\rangle \mapsto |\psi(t)\rangle$, called the time-evolution operator, is unitary. Unitarity of $U(t)$ guarantees that the norm of $|\psi(t)\rangle$ is conserved: $\langle\psi(0)|U^\dagger(t)U(t)|\psi(0)\rangle = \langle\psi(0)|\psi(0)\rangle = 1 \quad (\forall t > 0)$.

Two mutually commuting operators A and B have simultaneous eigenstates. If, in contrast, they do not commute, the measurement outcomes of these operators on any state $|\psi\rangle$ satisfy the following uncertainty relations. Let $\langle A \rangle = \langle \psi|A|\psi \rangle$ and $\langle B \rangle = \langle \psi|B|\psi \rangle$ be their respective expectation values and $\Delta A = \sqrt{\langle (A - \langle A \rangle)^2 \rangle}$ and $\Delta B = \sqrt{\langle (B - \langle B \rangle)^2 \rangle}$ be respective standard deviations. Then they satisfy

$$\Delta A \Delta B \geq \frac{1}{2}|\langle \psi|[A, B]|\psi \rangle|. \tag{7}$$

2.3. Simple example

Examples to clarify the axioms introduced in the previous subsection are given. They are used to controll quantum states in physical realizations of a quantum computer. A spin-1/2 particle has two states, which we call spin-up state $|\uparrow\rangle$ and spin-down state $|\downarrow\rangle$. It is common to assign components $|\uparrow\rangle = (1,0)^t$ and $|\downarrow\rangle = (0,1)^t$. They form a basis of a vector space $\mathbb{C}^2$.

Let us consider a time-independent Hamiltonian

$$H = -\frac{\hbar}{2}\omega\sigma_x \tag{8}$$

acting on the spin Hilbert space $\mathbb{C}^2$. Suppose the system is in the eigenstate of σ_z with the eigenvalue $+1$ at time $t = 0$; $|\psi(0)\rangle = |\uparrow\rangle$. The wave function $|\psi(t)\rangle$ $(t > 0)$ is then found from Eq. (5) as

$$|\psi(t)\rangle = \exp\left(i\frac{\omega}{2}\sigma_x t\right)|\psi(0)\rangle = \begin{pmatrix} \cos\omega t/2 & i\sin\omega t/2 \\ i\sin\omega t/2 & \cos\omega t/2 \end{pmatrix}\begin{pmatrix} 1 \\ 0 \end{pmatrix}$$

$$= \begin{pmatrix} \cos\omega t/2 \\ i\sin\omega t/2 \end{pmatrix} = \cos\frac{\omega}{2}t|\uparrow\rangle + i\sin\frac{\omega}{2}t|\downarrow\rangle. \tag{9}$$

Suppose we measure σ_z in $|\psi(t)\rangle$. The spin is found spin-up with probability $P_\uparrow(t) = \cos^2(\omega t/2)$ and spin-down with probability $P_\downarrow(t) = \sin^2(\omega t/2)$.

Consider a more general Hamiltonian

$$H = -\frac{\hbar}{2}\omega\hat{\boldsymbol{n}} \cdot \boldsymbol{\sigma}, \tag{10}$$

where $\hat{\boldsymbol{n}}$ is a unit vector in $\mathbb{R}^3$. The time-evolution operator is readily obtained, by making use of a well known formula

$$e^{i\alpha(\hat{\boldsymbol{n}}\cdot\boldsymbol{\sigma})} = \cos\alpha I + i(\hat{\boldsymbol{n}} \cdot \boldsymbol{\sigma})\sin\alpha \tag{11}$$

as

$$U(t) = \exp(-iHt/\hbar) = \cos\omega t/2\, I + i(\hat{\boldsymbol{n}} \cdot \boldsymbol{\sigma})\sin\omega t/2. \tag{12}$$

Suppose the initial state is $|\psi(0)\rangle = (1,0)^t$ for example. Then we find, at a later time $t > 0$,

$$|\psi(t)\rangle = U(t)|\psi(0)\rangle = \begin{pmatrix} \cos(\omega t/2) + in_z \sin(\omega t/2) \\ i(n_x + in_y)\sin(\omega t/2) \end{pmatrix}. \tag{13}$$

2.4. Multipartite system, tensor product and entangled state

So far, we have implictly assumed that the system is made of a single component. Suppose a system is made of two components, one lives in a Hilbert space $\mathcal{H}_1$ and the other in $\mathcal{H}_2$. A system composed of two separate components is called bipartite. The system as a whole lives in a Hilbert space $\mathcal{H} = \mathcal{H}_1 \otimes \mathcal{H}_2$, whose general vector is written as

$$|\psi\rangle = \sum_{i,j} c_{ij}|e_{1,i}\rangle \otimes |e_{2,j}\rangle, \tag{14}$$

where $\{|e_{a,i}\rangle\}$ $(a = 1, 2)$ is an orthonormal basis in $\mathcal{H}_a$ and $\sum_{i,j}|c_{ij}|^2 = 1$.

A state $|\psi\rangle \in \mathcal{H}$ written as a tensor product of two vectors as $|\psi\rangle = |\psi_1\rangle \otimes |\psi_2\rangle$, $(|\psi_a\rangle \in \mathcal{H}_a)$ is called a separable state or a tensor product state. A separable state admits a classical interpretation "The first system is in the state $|\psi_1\rangle$ while the second system is in $|\psi_2\rangle$". It is clear that the set of separable state has dimension $\dim \mathcal{H}_1 + \dim \mathcal{H}_2$. Note, however, that the total space $\mathcal{H}$ has different dimension than this: $\dim \mathcal{H} = \dim \mathcal{H}_1 \dim \mathcal{H}_2$. This number is considerably larger than the dimension of the sparable states when $\dim \mathcal{H}_a$ $(a = 1, 2)$ are large. What are the missing states then? Let us consider a spin state

$$|\psi\rangle = \frac{1}{\sqrt{2}}(|\uparrow\rangle \otimes |\uparrow\rangle + |\downarrow\rangle \otimes |\downarrow\rangle) \tag{15}$$

of two electrons. Suppose $|\psi\rangle$ may be decomposed as

$$|\psi\rangle = (c_1|\uparrow\rangle + c_2|\downarrow\rangle) \otimes (d_1|\uparrow\rangle + d_2|\downarrow\rangle)$$
$$= c_1 d_1|\uparrow\rangle \otimes |\uparrow\rangle + c_1 d_2|\uparrow\rangle \otimes |\downarrow\rangle + c_2 d_1|\downarrow\rangle \otimes |\uparrow\rangle + c_2 d_2|\downarrow\rangle \otimes |\downarrow\rangle.$$

However this decomposition is not possible since we must have $c_1 d_2 = c_2 d_1 = 0$, $c_1 d_1 = c_2 d_2 = 1/\sqrt{2}$ simultaneously and it is clear that the above equations have no common solution, showing $|\psi\rangle$ is not separable.

Such non-separable states are called entangled. Entangled states refuse classical descriptions. Entanglement is used extensively as a powerful computational resource in the following.

Suppose a bipartite state (14) is given. We are interested in when the state is separable and when entangled. The criterion is given by the Schmidt decomposition of $|\psi\rangle$.

Theorem 2.1. *Let $\mathcal{H} = \mathcal{H}_1 \otimes \mathcal{H}_2$ be the Hilbert space of a bipartite system. Then a vector $|\psi\rangle \in \mathcal{H}$ admits the Schmidt decomposition*

$$|\psi\rangle = \sum_{i=1}^{r} \sqrt{s_i} |f_{1,i}\rangle \otimes |f_{2,i}\rangle, \tag{16}$$

where $s_i > 0$ are called the Schmidt coefficients satisfying $\sum_i s_i = 1$ and $\{|f_{a,i}\rangle\}$ is an orthonormal set of $\mathcal{H}_a$. The number $r \in \mathbb{N}$ is called the Schmidt number of $|\psi\rangle$.

It follows from the above theorem that a bipartite state $|\psi\rangle$ is separable if and only if its Schmidt number r is 1. See[1] for the proof.

2.5. *Mixed states and density matrices*

It might happen in some cases that a quantum system under considertation is in the state $|\psi_i\rangle$ with a probability p_i. In other words, we cannot say definitely which state the system is in. Therefore some random nature comes into the description of the system. Such a system is said to be in a mixed state while a system whose vector is uniquely specified is in a pure state. A pure state is a special case of a mixed state in which $p_i = 1$ for some i and $p_j = 0$ $(j \neq i)$.

A particular state $|\psi_i\rangle \in \mathcal{H}$ appears with probability p_i in an ensemble of a mixed state, in which case the expectation value of the observable a is $\langle \psi_i | A | \psi_i \rangle$. The mean value of a averaged over the ensemble is then given by

$$\langle A \rangle = \sum_{i=1}^{N} p_i \langle \psi_i | A | \psi_i \rangle, \tag{17}$$

where N is the number of available states. Let us introduce the density matrix by

$$\rho - \sum_{i=1}^{N} p_i |\psi_i\rangle\langle\psi_i|. \tag{18}$$

Then Eq. (17) is rewritten in a compact form as $\langle A \rangle = \mathrm{Tr}(\rho A)$.

Let A be a Hermitian matrix. A is called positive-semidefinite if $\langle \psi | A | \psi \rangle \geq 0$ for any $|\psi\rangle \in \mathcal{H}$. It is easy to show all the eigenvalues of a positive-semidefinite Hermitian matrix are non-negative. Conversely, a Hermitian matrix A whose every eigenvalue is non-negative is positive-semidefinite.

Properties which a density matrix ρ satisfies are very much like axioms for pure states.

A 1' A physical state of a system, whose Hilbert space is $\mathcal{H}$, is completely specified by its associated density matrix $\rho : \mathcal{H} \to \mathcal{H}$. A density matrix is a positive-semidefinite Hermitian operator with $\operatorname{tr} \rho = 1$, see remarks below.

A 2' The mean value of an observable a is given by

$$\langle A \rangle = \operatorname{tr}(\rho A). \tag{19}$$

A 3' The temporal evolution of the density matrix follows the Liouville-von Neumann equation

$$i\hbar \frac{d}{dt} \rho = [H, \rho] \tag{20}$$

where H is the system Hamiltonian, see remarks below.

Several remarks are in order.

- The density matrix (18) is Hermitian since $p_i \in \mathbb{R}$. It is positive-semidefinite since $\langle \psi | \rho | \psi \rangle = \sum_i p_i |\langle \psi_i | \psi \rangle|^2 \geq 0$.
- Each $|\psi_i\rangle$ follows the Schrödinger equation $i\hbar \frac{d}{dt}|\psi_i\rangle = H|\psi_i\rangle$ in a closed quantum system. Its Hermitian conjugate is $-i\hbar \frac{d}{dt}\langle \psi_i| = \langle \psi_i|H$. We prove the Liouville-von Neumann equation from these equations as

$$i\hbar \frac{d}{dt}\rho = i\hbar \frac{d}{dt} \sum_i p_i |\psi_i\rangle\langle \psi_i| = \sum_i p_i H|\psi_i\rangle\langle \psi_i| - \sum_i p_i |\psi_i\rangle\langle \psi_i|H = [H, \rho].$$

We denote the set of all possible density matrices as $\mathcal{S}(\mathcal{H})$.

Example 2.1. A pure state $|\psi\rangle$ is a special case in which the corresponding density matrix is $\rho = |\psi\rangle\langle\psi|$. Therefore ρ is nothing but the projection operator onto the state. Observe that $\langle A \rangle = \operatorname{tr} \rho A = \sum_i \langle e_i|\psi\rangle\langle\psi|A|e_i\rangle = \langle \psi|A \sum_i |e_i\rangle\langle e_i|\psi\rangle = \langle \psi|A|\psi\rangle$, where $\{|e_i\rangle\}$ is an orthonormal set.

Let us consider a beam of photons. We take a horizontally polarized state $|e_1\rangle = | \leftrightarrow \rangle$ and a vertically polarized state $|e_2\rangle = | \updownarrow \rangle$ as orthonormal basis vectors. If the photons are a totally uniform mixture of two polarized states, the density matrix is given by

$$\rho = \frac{1}{2}|e_1\rangle\langle e_1| + \frac{1}{2}|e_2\rangle\langle e_2| = \frac{1}{2}\begin{pmatrix} 1 & 0 \\ 0 & 1 \end{pmatrix} = \frac{1}{2}I.$$

This state is called a maximally mixed state.

If photons are in a pure state $|\psi\rangle = (|e_1\rangle + |e_2\rangle)/\sqrt{2}$, the density matrix, with $\{|e_i\rangle\}$ as basis, is

$$\rho = |\psi\rangle\langle\psi| = \frac{1}{2}\begin{pmatrix} 1 & 1 \\ 1 & 1 \end{pmatrix}.$$

We are interested in when ρ represents a pure state or a mixed state.

Theorem 2.2. *A state ρ is pure if and only if $\operatorname{tr}\rho^2 = 1$.*

Proof: Since ρ is Hermitian, all its eigenvalues λ_i $(1 \le i \le \dim \mathcal{H})$ are real and the corresponding eigenvectors $\{|\lambda_i\rangle\}$ are made orthonormal. Then $\rho^2 = \sum_{i,j} \lambda_i\lambda_j|\lambda_i\rangle\langle\lambda_i|\lambda_j\rangle\langle\lambda_j| = \sum_i \lambda_i^2|\lambda_i\rangle\langle\lambda_i|$. Therefore $\operatorname{tr}\rho^2 = \sum_i \lambda_i^2 \le \lambda_{\max}\sum_i \lambda_i = \lambda_{\max} \le 1$, where $\lambda_{\max}$ is the largest eigenvalue of ρ. Therefore $\operatorname{tr}\rho^2 = 1$ implies $\lambda_{\max} = 1$ and all the other eigenvalues are zero. The converse is trivial. ∎

We classify mixed states into three classes, similarly to the classification of pure states into separable states and entangled states. We use a bipartite system in the definition but generalization to multipartitle systems should be obvious.

Definition 2.1. A state ρ is called separable if it is written in the form

$$\rho = \sum_i p_i \rho_{1,i} \otimes \rho_{2,i}, \tag{21}$$

where $0 \le p_i \le 1$ and $\sum_i p_i = 1$. It is called inseparable, if ρ does not admit the decomposition (21).

It is important to realize that only inseparable states have quantum correlations analogous to entangled pure states. It does not necessarily imply that a separable state has no non-classical correlation. It is pointed out that useful non-classical correlation exists in the subset of separable states.[11]

In the next subsection, we discuss how to find whether a given bipartite density matrix is separable or inseparable.

2.6. *Negativity*

Let ρ be a bipartite state and define the partial transpose ρ^{pt} of ρ with respect to the second Hilbert space as

$$\rho_{ij,kl} \rightarrow \rho_{il,kj}, \tag{22}$$

where $\rho_{ij,kl} = (\langle e_{1,i}| \otimes \langle e_{2,j}|) \, \rho \, (|e_{1,k}\rangle \otimes |e_{2,l}\rangle)$. Here $\{|e_{1,k}\rangle\}$ is the orthonormal basis of the first system while $\{|e_{2,k}\rangle\}$ of the second system. Suppose ρ takes a separable form (21). Then the partial transpose yields

$$\rho^{\text{pt}} = \sum_i p_i \rho_{1,i} \otimes \rho_{2,i}^t. \tag{23}$$

Note here that ρ^t for any density matrix ρ is again a density matrix since it is still positive semi-definite Hermitian with unit trace. Therefore the partial transposed density matrix (23) is another density matrix. It was conjectured by Peres[12] and subsequently proven by the Hordecki family[13] that positivity of the partially transposed density matrix is necessary and sufficient condition for ρ to be separable in the cases of $\mathbb{C}^2 \otimes \mathbb{C}^2$ systems and $\mathbb{C}^2 \otimes \mathbb{C}^3$ systems. Conversely, if the partial transpose of ρ of these systems is not a density matrix, then ρ is inseparable. Instead of giving the proof, we look at the following example.

Example 2.2. Let us consider the Werner state

$$\rho = \begin{pmatrix} \frac{1-p}{4} & 0 & 0 & 0 \\ 0 & \frac{1+p}{4} & -\frac{p}{2} & 0 \\ 0 & -\frac{p}{2} & \frac{1+p}{4} & 0 \\ 0 & 0 & 0 & \frac{1-p}{4} \end{pmatrix}, \tag{24}$$

where $0 \leq p \leq 1$. Here the basis vectors are arranged in the order

$$|e_{1,1}\rangle|e_{2,1}\rangle, |e_{1,1}\rangle|e_{2,2}\rangle, |e_{1,2}\rangle|e_{2,1}\rangle, |e_{1,2}\rangle|e_{2,2}\rangle.$$

Partial transpose of ρ yields

$$\rho^{\text{pt}} = \begin{pmatrix} \frac{1-p}{4} & 0 & 0 & -\frac{p}{2} \\ 0 & \frac{1+p}{4} & 0 & 0 \\ 0 & 0 & \frac{1+p}{4} & 0 \\ -\frac{p}{2} & 0 & 0 & \frac{1-p}{4} \end{pmatrix}.$$

ρ^{pt} must have non-negative eigenvalues to be a physically acceptable state. The characteristic equation of ρ^{pt} is

$$D(\lambda) = \det(\rho^{\text{pt}} - \lambda I) = \left(\lambda - \frac{p+1}{4}\right)^3 \left(\lambda - \frac{1-3p}{4}\right) = 0.$$

There are threefold degenerate eigenvalue $\lambda = (1+p)/4$ and nondegenerate eigenvalue $\lambda = (1-3p)/4$. This shows that ρ^{pt} is an unphysical state for $1/3 < p \leq 1$. If this is the case, ρ is inseparable.

From the above observation, entangled states are characterized by non-vanishing negativity defined as

$$N(\rho) \equiv \frac{1}{2}(\sum_i |\lambda_i| - 1). \tag{25}$$

Note that negativity vanishes if and only if all the eigenvalues of ρ^{pt} are nonnegative. However there is a class of inseparable states which are not characterized by negativity.

2.7. Partial trace and purification

Let $\mathcal{H} = \mathcal{H}_1 \otimes \mathcal{H}_2$ be a Hilbert space of a bipartite system made of components 1 and 2 and let A be an arbitrary operator acting on $\mathcal{H}$. The partial trace of A over $\mathcal{H}_2$ generates an operator acting on $\mathcal{H}_1$ defined as

$$A_1 = \text{tr}_2 A \equiv \sum_k (I \otimes \langle k|)A(I \otimes |k\rangle). \tag{26}$$

We will be concerned with the partial trace of a density matrix in practical applications. Let $\rho = |\psi\rangle\langle\psi| \in \mathcal{S}(\mathcal{H})$ be a density matrix of a pure state $|\psi\rangle$. Suppose we are interested only in the first system and have no access to the second system. Then the partial trace allows us to "forget" about the second system. In other words, the partial trace quantifies our ignarance on the second sytem.

To be concrete, consider a pure state

$$|\psi\rangle = \frac{1}{\sqrt{2}}(|e_1\rangle|e_1\rangle + |e_2\rangle|e_2\rangle),$$

where $\{|e_i\rangle\}$ is an orthonormal basis of $\mathbb{C}^2$. The corresponding density matrix is

$$\rho = \frac{1}{2}\begin{pmatrix} 1 & 0 & 0 & 1 \\ 0 & 0 & 0 & 0 \\ 0 & 0 & 0 & 0 \\ 1 & 0 & 0 & 1 \end{pmatrix},$$

where the basis vectors are ordered as $\{|e_1\rangle|e_1\rangle, |e_1\rangle|e_2\rangle, |e_2\rangle|e_1\rangle, |e_2\rangle|e_2\rangle\}$. The partial trace of ρ is

$$\rho_1 = \text{tr}_2\rho = \sum_{i=1,2} (I \otimes \langle e_i|)\rho(I \otimes |e_i\rangle) = \frac{1}{2}\begin{pmatrix} 1 & 0 \\ 0 & 1 \end{pmatrix}. \tag{27}$$

Note that a pure state $|\psi\rangle$ is mapped to a maximally mixed state ρ_1.

We have seen above that the partial trace of a pure-state density matrix of a bipartite system over one of the constituent Hilbert spaces yields a mixed state. How about the converse? Given a mixed state density matrix, is it always possible to find a pure state density matrix whose partial trace over the extra Hilbert space yields the given density matrix? The answer is yes and the process to find the pure state is called the purification. Let $\rho_1 = \sum_k p_k |\psi_k\rangle\langle\psi_k|$ be a general density matrix of a system 1 with the Hilbert space $\mathcal{H}_1$. Now let us introduce the second Hilbert space $\mathcal{H}_2$ whose dimension is the same as that of $\mathcal{H}_1$. Then formally introduce a normalized vector

$$|\Psi\rangle = \sum_k \sqrt{p_k} |\psi_k\rangle \otimes |\phi_k\rangle, \qquad (28)$$

where $\{|\phi_k\rangle\}$ is an orthonormal basis of $\mathcal{H}_2$. We find

$$\mathrm{tr}_2 |\Psi\rangle\langle\Psi| = \sum_{i,j,k} (I \otimes \langle\phi_i|) \left[\sqrt{p_j p_k} |\psi_j\rangle |\phi_j\rangle\langle\psi_k|\langle\phi_k| \right] (I \otimes |\phi_i\rangle)$$

$$= \sum_k p_k |\psi_k\rangle\langle\psi_k| = \rho_1. \qquad (29)$$

It is always possible to purify a mixed state by tensoring an extra Hilbert space of the same dimension as that of the original Hilbert space. Purification is far from unique.

3. Qubits

A (Boolean) bit assumes two distinct values, 0 and 1, and it constitutes the building block of the classical information theory. Quantum information theory, on the other hand, is based on qubits.

3.1. *One qubit*

A qubit is a (unit) vector in the vector space $\mathbb{C}^2$, whose basis vectors are denoted as

$$|0\rangle = (1,0)^t \text{ and } |1\rangle = (0,1)^t. \qquad (30)$$

What these vectors physically mean depends on the physical realization employed for quantum information processing.

They might represent spin states of an electron, $|0\rangle = |\uparrow\rangle$ and $|1\rangle = |\downarrow\rangle$. Electrons are replaced by nuclei with spin $1/2$ in NMR (Nuclear Magnetic Resonance).

In some cases, $|0\rangle$ stands for a vertically polarized photon $|\updownarrow\rangle$ while $|1\rangle$ represents a horizontally polarized photon $|\leftrightarrow\rangle$. Alternatively they might correspond to photons polarized in different directions. For example, $|0\rangle$ may represent a polarization state $|\nearrow\rangle = \frac{1}{\sqrt{2}}(|\updownarrow\rangle + |\leftrightarrow\rangle)$ while $|1\rangle$ represents a state $|\searrow\rangle = \frac{1}{\sqrt{2}}(|\updownarrow\rangle - |\leftrightarrow\rangle)$.

Truncated two states from many levels may be employed as a qubit. We may assign $|0\rangle$ to the ground state and $|1\rangle$ to the first excited state of an atom or an ion.

In any case, we have to fix a set of basis vectors when we carry out quantum information processing. In the following, the basis is written in an abstract form as $\{|0\rangle, |1\rangle\}$, unless otherwise stated.

It is convenient to assume the vector $|0\rangle$ corresponds to the classical bit 0, while $|1\rangle$ to 1. Moreover a qubit may be in a superposition state:

$$|\psi\rangle = a|0\rangle + b|1\rangle, \quad \text{with} \quad |a|^2 + |b|^2 = 1. \tag{31}$$

If we measure $|\psi\rangle$ to see whether it is in $|0\rangle$ or $|1\rangle$, the outcome will be 0 (1) with the probability $|a|^2$ ($|b|^2$) and the state immediately after the measurement is $|0\rangle$ ($|1\rangle$).

Although a qubit may take infinitely many different states, it should be kept in mind that we can extract from it as the same amount of information as that of a classical bit. Information can be extracted only through measurements. When we measure a qubit, the state vector 'collapses' to the eigenvector that corresponds to the eigenvalue observed. Suppose a spin is in the state $a|0\rangle + b|1\rangle$. If we observe that the z-component of the spin is $+1/2$, the system immediately after the measurement is in $|0\rangle$. This happens with probability $\langle\psi|0\rangle\langle0|\psi\rangle = |a|^2$. The measurement outcome of a qubit is always one of the eigenvalues, which we call abstractly 0 and 1.

3.2. Bloch sphere

It is useful, for many purposes, to express a state of a single qubit graphically. Let us parameterize a one-qubit pure state $|\psi\rangle$ with θ and ϕ as

$$|\psi(\theta, \phi)\rangle = \cos\frac{\theta}{2}|0\rangle + e^{i\phi}\sin\frac{\theta}{2}|1\rangle. \tag{32}$$

The phase of $|\psi\rangle$ is fixed in such a way that the coefficient of $|0\rangle$ is real. It is easy to verify that $(\hat{\boldsymbol{n}}(\theta, \phi) \cdot \boldsymbol{\sigma})|\psi(\theta, \phi)\rangle = |\psi(\theta, \phi)\rangle$, where $\boldsymbol{\sigma} = (\sigma_x, \sigma_y, \sigma_z)$ and $\hat{\boldsymbol{n}}(\theta, \phi)$ is a real unit vector called the Bloch vector with components $\hat{\boldsymbol{n}}(\theta, \phi) = (\sin\theta\cos\phi, \sin\theta\sin\phi, \cos\theta)^t$. It is therefore natural to assign $\hat{\boldsymbol{n}}(\theta, \phi)$ to a state vector $|\psi(\theta, \phi)\rangle$ so that $|\psi(\theta, \phi)\rangle$ is expressed as a unit vector $\hat{\boldsymbol{n}}(\theta, \phi)$ on the surface of the unit sphere, called the Bloch sphere. This correspondence is one-to-one if the ranges of θ and ϕ are restricted to $0 \leq \theta \leq \pi$ and $0 \leq \phi < 2\pi$.

It is verified that state (32) satisfies

$$\langle \psi(\theta, \phi) | \boldsymbol{\sigma} | \psi(\theta, \phi) \rangle = \hat{\boldsymbol{n}}(\theta, \phi). \tag{33}$$

A density matrix ρ of a qubit can be represented as a point on a unit ball. Since ρ is a positive semi-definite Hermitian matrix with unit trace, its most general form is

$$\rho = \frac{1}{2} \left(I + \sum_{i=x,y,z} u_i \sigma_i \right), \tag{34}$$

where $\vec{u} \in \mathbb{R}^3$ satisfies $|\boldsymbol{u}| \leq 1$. The reality follows from the Hermiticity requirement and $\text{tr}\,\rho = 1$ is obvious. The eigenvalues of ρ are $\lambda_{\pm} = \frac{1}{2}\left(1 \pm \sqrt{|\boldsymbol{u}|}\right)/2$ and therefore non-negative. The eigenvalue λ_- vanishes in case $|\boldsymbol{u}| = 1$, for which rank $\rho = 1$. Therefore the surface of the unit sphere corresponds to pure states. The converse is also shown easily. In contrast, all the points $\boldsymbol{u}$ inside a unit ball correspond to mixed states. The ball is called the Bloch ball and the vector $\boldsymbol{u}$ is also called the Bloch vector.

It is easily verified that ρ given by Eq. (34) satisfies

$$\langle \boldsymbol{\sigma} \rangle = \text{tr}\,(\rho \boldsymbol{\sigma}) = \boldsymbol{u}. \tag{35}$$

3.3. Multi-qubit systems and entangled states

Let us consider a group of many (n) qubits next. Such a system behaves quite differently from a classical one and this difference gives a distinguishing aspect to quantum information theory. An n-qubit system is often called a (quantum) register in the context of quantum computing.

As an example, let us consider an n-qubit register. Suppose we specify the state of each qubit separately like a classical case. Each of the qubit is then described by a 2-d complex vector of the form $a_i|0\rangle + b_i|1\rangle$ and we need $2n$ complex numbers $\{a_i, b_i\}_{1 \leq i \leq n}$ to specify the state. This corresponds the a tensor product state $(a_1|0\rangle + b_1|1\rangle) \otimes \ldots \otimes (a_n|0\rangle + b_n|1\rangle) \in \mathbb{C}^{2n}$. If the system is treated in a fully quantum-mechanical way, however, a general state vector of the register is represented as

$$|\psi\rangle = \sum_{i_k=0,1} a_{i_1 i_2 \ldots i_n} |i_1\rangle \otimes |i_2\rangle \otimes \ldots \otimes |i_n\rangle \in \mathbb{C}^{2^n}.$$

Note that $2^n \gg 2n$ for a large number n. The ratio $2^n/2n$ is $\sim 10^{298}$ for $n = 1000$. Most quantum states in a Hilbert space with large n are entangled having no classical analogues. Entanglement is an extremely powerful resource for quantum computation and quantum communication.

Let us consider a 2-qubit system for definiteness. The system has a binary basis $\{|00\rangle, |01\rangle, |10\rangle, |11\rangle\}$. More generally, a basis for a system of n qubits may be $\{|b_{n-1}b_{n-2}\ldots b_0\rangle\}$, where $b_{n-1}, b_{n-2}, \ldots, b_0 \in \{0, 1\}$. It is also possible to express the basis in terms of the decimal system. We write $|x\rangle$, instead of $|b_{n-1}b_{n-2}\ldots b_0\rangle$, where $x = b_{n-1}2^{n-1} + b_{n-2}2^{n-2} + \ldots + b_0$. The basis for a 2-qubit system may be written also as $\{|0\rangle, |1\rangle, |2\rangle, |3\rangle\}$ with this decimal notation.

The set

$$\{|\Phi^+\rangle = \frac{1}{\sqrt{2}}(|00\rangle + |11\rangle), \quad |\Phi^-\rangle = \frac{1}{\sqrt{2}}(|00\rangle - |11\rangle),$$

$$|\Psi^+\rangle = \frac{1}{\sqrt{2}}(|01\rangle + |10\rangle), \quad |\Psi^-\rangle = \frac{1}{\sqrt{2}}(|01\rangle - |10\rangle)\} \tag{36}$$

is an orthonormal basis of a two-qubit system and is called the Bell basis. Each vector is called the Bell state or the Bell vector. Note that all the Bell states are entangled.

4. Quantum Gates, Quantum Circuit and Quantum Computation

4.1. *Introduction*

Now that we have introduced qubits to store information, it is time to consider operations acting on them. If they are simple, these operations are called gates, or quantum gates, in analogy with those in classical logic circuits. More complicated quantum circuits are composed of these simple gates. A collection of quantum circuits for executing a complicated algorithm, a quantum algorithm, is a part of a quantum computation.

Definition 4.1. (Quantum Computation) A quantum computation is a collection of the following three elements:

(1) A register or a set of registers,
(2) A unitary matrix u, which is taylored to execute a given quantum algorithm and
(3) Measurements to extract information we need.

More formally, a quantum computation is the set $\{\mathcal{H}, U, \{M_m\}\}$, where $\mathcal{H} = \mathbb{C}^{2^n}$ is the Hilbert space of an n-qubit register, $U \in U(2^n)$ represents a quantum algorithm and $\{M_m\}$ is the set of measurement operators. The hardware (1) is called a quantum computer.

Suppose the register is set to a fiducial initial state, $|\psi_{\text{in}}\rangle = |00\ldots0\rangle$ for example. A unitary matrix U_{alg} is generated by an algorithm which we want to execute. Operation of U_{alg} on $|\psi_{\text{in}}\rangle$ yields the output state $|\psi_{\text{out}}\rangle = U_{\text{alg}}|\psi_{\text{in}}\rangle$. Information is extracted from $|\psi_{\text{out}}\rangle$ by appropriate measurements.

4.2. Quantum gates

We have so far studied the change of a state upon measurements. When measurements are not made, the time evolution of a state is described by the Schrödinger equation. The time evolution operator U is unitary: $UU^{\dagger} = U^{\dagger}U = I$. We will be free from the Schrödinger equation in the following and assume there always exist unitary matrices which we need.

One of the important conclusions derived from the unitarity of gates is that the computational process is reversible.

4.2.1. Simple quantum gates

Examples of quantum gates which transform a one-qubit state are given below. We call them one-qubit gates in the following. Linearity guarantees that the action of a gate is completely specified if its action on the basis $\{|0\rangle, |1\rangle\}$ is given. Consider the gate I whose action on the basis vectors is $I : |0\rangle \to |0\rangle, |1\rangle \to |1\rangle$. The matrix expression of this gate is

$$I = |0\rangle\langle0| + |1\rangle\langle1| = \begin{pmatrix} 1 & 0 \\ 0 & 1 \end{pmatrix}. \tag{37}$$

Similarly we introduce $X : |0\rangle \to |1\rangle, |1\rangle \to |0\rangle, Y : |0\rangle \to -|1\rangle, |1\rangle \to |0\rangle$ and $Z : |0\rangle \to |0\rangle, |1\rangle \to -|1\rangle$ by

$$X = |1\rangle\langle0| + |0\rangle\langle1| = \begin{pmatrix} 0 & 1 \\ 1 & 0 \end{pmatrix} = \sigma_x, \tag{38}$$

$$Y = |0\rangle\langle1| - |1\rangle\langle0| = \begin{pmatrix} 0 & -1 \\ 1 & 0 \end{pmatrix} = -i\sigma_y, \tag{39}$$

$$Z = |0\rangle\langle0| - |1\rangle\langle1| = \begin{pmatrix} 1 & 0 \\ 0 & -1 \end{pmatrix} = \sigma_z. \tag{40}$$

The transformation I is the identity transformation, while X is the negation (NOT), Z the phase shift and $Y = XZ$ the combination thereof.

CNOT (controlled-NOT) gate is a 2-qubit gate, which plays an important role. The gate flips the second qubit (the target qubit) when the first qubit (the control qubit) is $|1\rangle$, while leaving the second bit unchanged when the first bit is $|0\rangle$. Let $\{|00\rangle, |01\rangle, |10\rangle, |11\rangle\}$ be a basis for the 2-qubit system. We use the standard basis vectors with components

$$|00\rangle = (1,0,0,0)^t, \ |01\rangle = (0,1,0,0)^t, \ |10\rangle = (0,0,1,0)^t, \ |11\rangle = (0,0,0,1)^t.$$

The action of CNOT gate, whose matrix expression will be written as U_{CNOT}, is $U_{\text{CNOT}} : |00\rangle \mapsto |00\rangle$, $|01\rangle \mapsto |01\rangle$, $|10\rangle \mapsto |11\rangle$, $|11\rangle \mapsto |10\rangle$. It has two equivalent expressions

$$\begin{aligned} U_{\text{CNOT}} &= |00\rangle\langle 00| + |01\rangle\langle 01| + |11\rangle\langle 10| + |10\rangle\langle 11| \\ &= |0\rangle\langle 0| \otimes I + |1\rangle\langle 1| \otimes X, \end{aligned} \tag{41}$$

having a matrix form

$$U_{\text{CNOT}} = \begin{pmatrix} 1 & 0 & 0 & 0 \\ 0 & 1 & 0 & 0 \\ 0 & 0 & 0 & 1 \\ 0 & 0 & 1 & 0 \end{pmatrix}. \tag{42}$$

Let $\{|i\rangle\}$ be the basis vectors, where $i \in \{0,1\}$. The action of CNOT on the input state $|i,j\rangle$ is written as $|i, i \oplus j\rangle$, where $i \oplus j$ is an addition $\mod 2$.

A 1-qubit gate whose unitary matrix is U is graphically depicted as

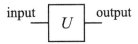

The left horizontal line is the input qubit while the right horizontal line is the output qubit: time flows from the left to the right.

A CNOT gate is expressed as

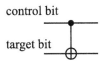

where $\bullet$ denotes the control bit, while $\oplus$ denotes the conditional negation. There may be many control bits (see CCNOT gate below). More generally, we consider a controlled-U gate, $V = |0\rangle\langle 0| \otimes I + |1\rangle\langle 1| \otimes U$, in which the target bit is acted on by a unitary transformation U only when the control bit is $|1\rangle$. This gate is denoted graphically as control bit

target bit

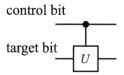

CCNOT (Controlled-Controlled-NOT) gate has three inputs and the third qubit flips only when the first two qubits are both in the state $|1\rangle$. The explicit form of the CCNOT gate is

$$U_{\text{CCNOT}} = (|00\rangle\langle 00| + |01\rangle\langle 01| + |10\rangle\langle 10|) \otimes I + |11\rangle\langle 11| \otimes X. \quad (43)$$

This gate is graphically expressed as

control bit 1

control bit 2

target bit

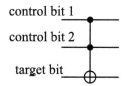

4.2.2. Walsh-Hadamard transformation

The Hadamard gate or the Hadamard transformation H is an important unitary transformation defined by

$$
\begin{aligned}
U_{\text{H}} : \ |0\rangle &\rightarrow \frac{1}{\sqrt{2}}(|0\rangle + |1\rangle) \\
: \ |1\rangle &\rightarrow \frac{1}{\sqrt{2}}(|0\rangle - |1\rangle).
\end{aligned}
\quad (44)
$$

The matrix representation of H is

$$U_{\text{H}} = \frac{1}{\sqrt{2}}(|0\rangle + |1\rangle)\langle 0| + \frac{1}{\sqrt{2}}(|0\rangle - |1\rangle)\langle 1| = \frac{1}{\sqrt{2}} \begin{pmatrix} 1 & 1 \\ 1 & -1 \end{pmatrix}. \quad (45)$$

A Hadamard gate is depicted as

$$-\boxed{H}-$$

There are numerous important applications of the Hadamard transformation. All possible 2^n states are generated when U_{H} is applied on each qubit of the state $|00\ldots0\rangle$:

$$(U_{\mathrm{H}} \otimes U_{\mathrm{H}} \otimes \ldots \otimes U_{\mathrm{H}})|00\ldots0\rangle$$

$$= \frac{1}{\sqrt{2}}(|0\rangle + |1\rangle) \otimes \frac{1}{\sqrt{2}}(|0\rangle + |1\rangle) \otimes \ldots \frac{1}{\sqrt{2}}(|0\rangle + |1\rangle)$$

$$= \frac{1}{\sqrt{2^n}} \sum_{x=0}^{2^n-1} |x\rangle. \tag{46}$$

Therefore, we produce a superposition of all the states $|x\rangle$ with $0 \leq x \leq 2^n - 1$ simultaneously. The transformation $U_{\mathrm{H}}^{\otimes n}$ is called the Walsh transformation, or Walsh-Hadamard transformation and denoted as W_n.

The quantum circuit

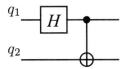

is used to generate Bell states from inputs $|00\rangle, |01\rangle, |10\rangle$ and $|11\rangle$.

4.2.3. SWAP gate and Fredkin gate

The SWAP gate acts on a tensor product state as

$$U_{\mathrm{SWAP}}|\psi_1, \psi_2\rangle = |\psi_2, \psi_1\rangle. \tag{47}$$

The explict form of U_{SWAP} is given by

$$U_{\mathrm{SWAP}} = |00\rangle\langle00| + |01\rangle\langle10| + |10\rangle\langle01| + |11\rangle\langle11| = \begin{pmatrix} 1 & 0 & 0 & 0 \\ 0 & 0 & 1 & 0 \\ 0 & 1 & 0 & 0 \\ 0 & 0 & 0 & 1 \end{pmatrix}. \tag{48}$$

The SWAP gate is expressed as

$$|\psi_1\rangle \;\rule{1cm}{0.4pt}\hspace{-0.2cm}\times\hspace{-0.2cm}\rule{1cm}{0.4pt}\; |\psi_2\rangle$$
$$|\psi_2\rangle \;\rule{1cm}{0.4pt}\hspace{-0.2cm}\times\hspace{-0.2cm}\rule{1cm}{0.4pt}\; |\psi_1\rangle$$

Note that the SWAP gate is a special gate which maps an arbitrary tensor product state to a tensor product state. In contrast, most 2-qubit gates map a tensor product state to an entangled state.

The controlled-SWAP gate

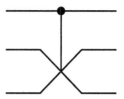

is also called the Fredkin gate. It flips the second (middle) and the third (bottom) qubits only when the first (top) qubit is in the state $|1\rangle$. Its explicit form is $U_{\text{Fredkin}} = |0\rangle\langle 0| \otimes I_4 + |1\rangle\langle 1| \otimes U_{\text{SWAP}}$.

4.3. No-cloning theorem

Theorem 4.1. *(Wootters and Zurek[14]) An unknown quantum system cannot be cloned by unitary transformations.*

Proof: Suppose there would exist a unitary transformation U that makes a clone of a quantum system. Namely, suppose U acts, for any state $|\varphi\rangle$, as $U : |\varphi 0\rangle \to |\varphi\varphi\rangle$. Let $|\varphi\rangle$ and $|\phi\rangle$ be two states that are linearly independent. Then we should have $U|\varphi 0\rangle = |\varphi\varphi\rangle$ and $U|\phi 0\rangle = |\phi\phi\rangle$ by definition. Then the action of U on $|\psi\rangle = \dfrac{1}{\sqrt{2}}(|\varphi\rangle + |\phi\rangle)$ yields

$$U|\psi 0\rangle = \frac{1}{\sqrt{2}}(U|\varphi 0\rangle + U|\phi 0\rangle) = \frac{1}{\sqrt{2}}(|\varphi\varphi\rangle + |\phi\phi\rangle).$$

If U were a cloning transformation, we must also have

$$U|\psi 0\rangle = |\psi\psi\rangle = \frac{1}{2}(|\varphi\varphi\rangle + |\varphi\phi\rangle + |\phi\varphi\rangle + |\phi\phi\rangle),$$

which contradicts the previous result. Therefore, there does not exist a unitary cloning transformation. ∎

Note however that the theorem does not apply if the states to be cloned are limited to $|0\rangle$ and $|1\rangle$. For these cases, the copying operator U should work as $U : |00\rangle \mapsto |00\rangle$, $: |10\rangle \mapsto |11\rangle$. We can assign arbitrary action of U on a state whose second input is $|1\rangle$ since this case will never happen. What we have to keep in mind is only that U be unitary. An example of such U is

$$U = (|00\rangle\langle 00| + |11\rangle\langle 10|) + (|01\rangle\langle 01| + |10\rangle\langle 11|). \tag{49}$$

where the first set of operators renders U the cloning operator and the second set is added just to make U unitary. We immediately notice that U is nothing but the CNOT gate.

Therefore, if the data under consideration is limited within $|0\rangle$ and $|1\rangle$, we can copy the qubit states even in a quantum computer. This fact is used to construct quantum error correcting codes.

4.4. *Quantum teleportation*

The purpose of quantum teleportation is to transmit an unknown quantum *state* of a qubit using two classical bits in such a way that the recipient reproduces the same state as the original qubit state. Note that the qubit itself is not transported but the information required to reproduce the quantum state is transmitted. The original state is destroyed such that quantum teleportation is not in contradiction with the no-cloning theorem.

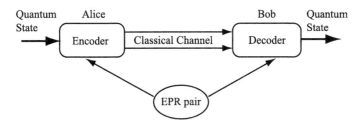

Fig. 1. In quantum teleportation, Alice sends Bob two classical bits so that Bob reproduces a qubit state Alice initially had.

Alice: Alice has a qubit, whose state she does *not* know. She wishes to send Bob the quantum state of this qubit through a classical communication channel. Let $|\phi\rangle = a|0\rangle + b|1\rangle$ be the state of the qubit. Both of them have been given one of the qubits of the entangled pair

$$|\Phi^+\rangle = \frac{1}{\sqrt{2}}(|00\rangle + |11\rangle)$$

in advance. They start with the state

$$|\phi\rangle \otimes |\Phi^+\rangle = \frac{1}{\sqrt{2}}\left(a|000\rangle + a|011\rangle + b|100\rangle + b|111\rangle\right), \qquad (50)$$

where Alice possesses the first two qubits while Bob has the third. Alice applies $U_{\mathrm{CNOT}} \otimes I$ followed by $U_{\mathrm{H}} \otimes I \otimes I$ to this state, which results in

$$(U_{\mathrm{H}} \otimes I \otimes I)(U_{\mathrm{CNOT}} \otimes I)(|\phi\rangle \otimes |\Phi^+\rangle)$$
$$= \frac{1}{2}[|00\rangle(a|0\rangle + b|1\rangle) + |01\rangle(a|1\rangle + b|0\rangle)$$
$$+ |10\rangle(a|0\rangle - b|1\rangle) + |11\rangle(a|1\rangle - b|0\rangle)]. \qquad (51)$$

If Alice measures the 2 qubits in her hand, she will obtain one of the states $|00\rangle, |01\rangle, |10\rangle$ or $|11\rangle$ with equal probability $1/4$. Bob's qubit (one of the EPR pair previously) collapses to $a|0\rangle + b|1\rangle, a|1\rangle + b|0\rangle, a|0\rangle - b|1\rangle$ or $a|1\rangle - b|0\rangle$, respectively, depending on the result of Alice's measurement. Alice then sends Bob her result of the measurement using two classical bits.

Bob: After receiving two classical bits, Bob knows the state of the qubit in his hand;

$$
\begin{array}{ccc}
\text{received bits} & \text{Bob's state} & \text{decoding} \\
00 & a|0\rangle + b|1\rangle & I \\
01 & a|1\rangle + b|0\rangle & X \\
10 & a|0\rangle - b|1\rangle & Z \\
11 & a|1\rangle - b|0\rangle & Y
\end{array}
\tag{52}
$$

Bob reconstructs the intial state $|\phi\rangle$ by applying the decoding process shown above. Suppose Alice sends Bob the classical bits 10, for example. Then Bob applies Z on his qubit to reconstruct $|\phi\rangle$ as $Z : (a|0\rangle - b|1\rangle) \mapsto (a|0\rangle + b|1\rangle) = |\phi\rangle$.

Figure 2 shows the actual quantum circuit for quantum teleportation.

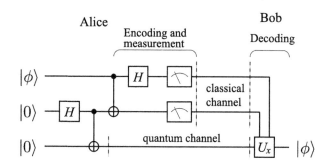

Fig. 2.   Quantum circuit implementation of quantum teleportation.

4.5. Universal quantum gates

It can be shown that any classical logic gate can be constructed by using a small set of gates, AND, NOT and XOR for example. Such a set of gates is called the *universal* set of gates. It can be shown that the CCNOT gate simulates these classical gates, and hence quantum circuits simulate any classical circuits. The set of quantum gates is, however, much larger than those classical gates. Thus we want to find a universal set of *quantum* gates from which any quantum circuits can be constructed.

It can be shown that
(1) the set of single qubit gates and
(2) CNOT gate
form a universal set of quantum circuits (universality theorem). The proof is highly technical and is not given here.[1,2,16] We, instead, sketch the proof in several lines.

It can be shown that any $U \in \mathrm{U}(n)$ is written as a product of N two-level unitary matrices, where $N \leq n(n-1)/2$ and a two-level unitary matrix is a unit matrix I_n in which only four components V_{aa}, V_{ab}, V_{ba} and V_{bb} are different from I_n. Moreover $V = (V_{ij})$ is an element of $\mathrm{U}(2)$. An example of a two-level unitary matrix is

$$V = \begin{pmatrix} \alpha^* & 0 & 0 & \beta^* \\ 0 & 1 & 0 & 0 \\ 0 & 0 & 1 & 0 \\ -\beta & 0 & 0 & \alpha \end{pmatrix}, \quad (|\alpha|^2 + |\beta|^2 = 1)$$

where $a = 1$ and $b = 4$.

Now we need to prove the universality theorem for two-level unitary matrices, which is certainly simpler than the general proof. By employing CNOT gates and their generalizations, it is possible to move the elements V_{aa}, V_{ab}, V_{ba} and V_{bb} so that they acts on a single qubit in the register. We need to implement the controlled-V gate whose target qubit is the one on which V acts. Implementation of the controlled-V gate requires generalized CNOT gates and several $\mathrm{U}(2)$ gates.[1,2,16]

4.6. *Quantum parallelism and entanglement*

Given an input x, a typical quantum computer "computes" $f(x)$ as

$$U_f : |x\rangle|0\rangle \mapsto |x\rangle|f(x)\rangle, \tag{53}$$

where U_f is a unitary matrix which implements the function f.

Suppose U_f acts on an input which is a superposition of many $|x\rangle$. Since U_f is a linear operator, it acts on all the constituent vectors of the superposition simultaneously. The output is also a superposition of all the results;

$$U_f : \sum_x |x\rangle|0\rangle \mapsto \sum_x |x\rangle|f(x)\rangle. \tag{54}$$

This feature, called the *quantum parallelism*, gives quantum computer an enormous power. A quantum computer is advantageous over a classical counterpart in that it makes use of this quantum parallelism and also entanglement.

A unitary transformation acts on a superposition of all possible states in most quantum algorithms. This superposition is prepared by the action of the Walsh-Hadamard transformation on an n-qubit register in the initial state $|00\ldots0\rangle = |0\rangle \otimes |0\rangle \otimes \ldots \otimes |0\rangle$ resulting in $\sum_{x=0}^{2^n-1} |x\rangle/\sqrt{2^n}$. This state is a superposition of vectors encoding all the integers between 0 and $2^n - 1$. Then the linearlity of U_f leads to

$$U_f \left(\frac{1}{\sqrt{2^n}} \sum_{x=0}^{2^n-1} |x\rangle|0\rangle \right) = \frac{1}{\sqrt{2^n}} \sum_{x=0}^{2^n-1} U_f|x\rangle|0\rangle = \frac{1}{\sqrt{2^n}} \sum_{x=0}^{2^n-1} |x\rangle|f(x)\rangle. \quad (55)$$

Note that the superposition is made of $2^n = e^{n\ln 2}$ states, which makes quantum computation exponentially faster than classical counterpart in a certain kind of computation.

What about the limitation of a quantum computer[3]? Let us consider the CCNOT gate for example. This gate flips the third qubit if and only if the first and the second qubits are both in the state $|1\rangle$ while it leaves the third qubit unchanged otherwise. Let us fix the third input qubit to $|0\rangle$. The third output qubit state is $|x \wedge y\rangle$, where $|x\rangle$ and $|y\rangle$ are the first and the second input qubits respectively. Suppose the input state of the first and the second qubits is a superposition of all possible states while the third qubit is fixed to $|0\rangle$. This can be achieved by the Walsh-Hadamard transformation as

$$U_H|0\rangle \otimes U_H|0\rangle \otimes |0\rangle = \frac{1}{\sqrt{2}}(|0\rangle + |1\rangle) \otimes \frac{1}{\sqrt{2}}(|0\rangle + |1\rangle) \otimes |0\rangle$$

$$= \frac{1}{2}(|000\rangle + |010\rangle + |100\rangle + |110\rangle). \quad (56)$$

By operating CCNOT on this state, we obtain

$$U_{CCNOT}(U_H|0\rangle \otimes U_H|0\rangle \otimes |0\rangle) = \frac{1}{2}(|000\rangle + |010\rangle + |100\rangle + |111\rangle). \quad (57)$$

This output may be thought of as the truth table of AND: $|x, y, x \wedge y\rangle$. It is extremely important to note that the output is an entangled state and the measurement projects the state to *one line* of the truth table, i.e., a single term in the RHS of Eq. (57).

There is no advantage of quantum computation over classical one at this stage. This is because only *one* result may be obtained by a single set of measurements. What is worse, we cannot choose a specific vector $|x, y, x \wedge y\rangle$ at our will! Thus any quantum algorithm should be programmed so that the particular vector we want to observe should have larger probability to be measured compared to other vectors. The programming strategies to deal with this feature are

(1) to amplify the amplitude, and hence the probability, of the vector that we want to observe. This strategy is employed in the Grover's database search algorithm.
(2) to find a common property of all the $f(x)$. This idea was employed in the quantum Fourier transform to find the order[a] of f in the Shor's factoring algorithm.

Now we consider the power of entanglement. Suppose we have an n-qubit register, whose Hilbert space is 2^n-dimensional. Since each qubit has two basis states $\{|0\rangle, |1\rangle\}$, there are $2n$ basis states, i.e., n $|0\rangle$'s and n $|1\rangle$'s, involved to span this Hilbert space. Imagine that we have a single quantum system, instead, which has the same Hilbert space. One might think that the system may do the same quantum computation as the n-qubit register does. One possible problem is that one cannot "measure the kth digit" leaving other digits unaffected. Even worse, consider how many different basis vectors are required for this system. This single system must have an enormous number, 2^n, of basis vectors! Multipartite implementation of a quantum algorithm requires exponentially smaller number of basis vectors than monopartite implementation since the former makes use of entanglement as a computational resource.

5. Simple Quantum Algorithms

Let us introduce a few simple quantum algorithms which will be of help to understand how quantum algorithms are different from and superior to classical algorithms.

5.1. *Deutsch algorithm*

The Deutsch algorithm is one of the first quantum algorithms which showed quantum algorithms may be more efficient than their classical counterparts.

---

[a]Let $m, N \in \mathbb{N}$ $(m < N)$ be numbers coprime to each other. Then there exists $P \in \mathbb{N}$ such that $m^P \equiv 1 \pmod{N}$. The smallest such number P is called the period or the order. It is easily seen that $m^{x+P} \equiv m^x \pmod{N}$, $\forall x \in \mathbb{N}$.

In spite of its simplicty, full usage of superposition principle and entanglement has been made here.

Let $f : \{0,1\} \rightarrow \{0,1\}$ be a binary function. Note that there are only four possible f, namely

$$f_1 : 0 \mapsto 0, \ 1 \mapsto 0, \quad f_2 : 0 \mapsto 1, \ 1 \mapsto 1,$$
$$f_3 : 0 \mapsto 0, \ 1 \mapsto 1, \quad f_4 : 0 \mapsto 1, \ 1 \mapsto 0.$$

First two cases, f_1 and f_2, are called *constant*,while the rest, f_3 and f_4, are *balanced*.If we only have classical resources, we need to evaluate f twice to tell if f is constant or balanced. There is a quantum algorithm, in contrast, with which it is possible to tell if f is constant or balanced with a single evaluation of f, as was shown by Deutsch.[18]

Let $|0\rangle$ and $|1\rangle$ correspond to classical bits 0 and 1, respectively, and consider the state $|\psi_0\rangle = \frac{1}{2}(|00\rangle - |01\rangle + |10\rangle - |11\rangle)$. We apply f on this state in terms of the unitary operator $U_f : |x,y\rangle \mapsto |x, y \oplus f(x)\rangle$, where $\oplus$ is an addition mod 2. To be explicit, we obtain

$$|\psi_1\rangle = U_f|\psi_0\rangle = \frac{1}{2}(|0, f(0)\rangle - |0, \neg f(0)\rangle + |1, f(1)\rangle - |1, \neg f(1)\rangle),$$

where $\neg$ stands for negation. Therefore this operation is nothing but the CNOT gate with the control bit $f(x)$; the target bit y is flipped if and only if $f(x) = 1$ and left unchanged otherwise. Subsequently we apply the Hadamard gate on the first qubit to obtain

$$|\psi_2\rangle = U_H|\psi_1\rangle$$
$$= \frac{1}{2\sqrt{2}}[(|0\rangle + |1\rangle)(|f(0)\rangle - |\neg f(0)\rangle) + (|0\rangle - |1\rangle)(|f(1)\rangle - |\neg f(1)\rangle)]$$

The wave function reduces to

$$|\psi_2\rangle = \frac{1}{\sqrt{2}}|0\rangle(|f(0)\rangle - |\neg f(0)\rangle) \tag{58}$$

in case f is constant, for which $|f(0)\rangle = |f(1)\rangle$, and

$$|\psi_2\rangle = \frac{1}{\sqrt{2}}|1\rangle(|f(0)\rangle - |f(1)\rangle) \tag{59}$$

if f is balanced, for which $|\neg f(0)\rangle = |f(1)\rangle$. Therefore the measurement of the first qubit tells us whether f is constant or balanced.

Let us consider a quantum circuit which implements the Deutsch algorithm. We first apply the Walsh-Hadamard transformation $W_2 = U_H \otimes U_H$ on $|01\rangle$ to obtain $|\psi_0\rangle$. We need to introduce a conditional gate U_f, i.e., the controlled-NOT gate with the control bit $f(x)$, whose action is

$U_f : |x, y\rangle \rightarrow |x, y \oplus f(x)\rangle$. Then the Hadamard gate is applied on the first qubit before it is measured. Figure 3 depicts this implementation.

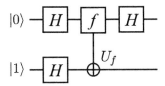

Fig. 3.   Implementation of the Deutsch algorithm.

In the quauntum circuit, we assume the gate U_f is a black box for which we do not ask the explicit implementation. We might think it is a kind of subroutine. Such a black box is often called an oracle. The gate U_f is called the Deutsch oracle. Its implementation is given only after f is specified.

Then what is the merit of the Deutsch algorithm? Suppose your friend gives you a unitary matrix U_f and asks you to tell if f is constant or balanced. Instead of applying $|0\rangle$ and $|1\rangle$ separately, you may contruct the circuit in Fig. 3 with the given matrix U_f and apply the circuit on the input state $|01\rangle$. Then you can tell your friend whether f is constant or balanced with a single use of U_f.

5.2. Deutsch-Jozsa algorithm

The Deutsch algorithm introduced in the previous section may be generalized to the Deutsch-Jozsa algorithm.[19] Let us first define the Deutsch-Jozsa problem. Suppose there is a binary function

$$f : S_n \equiv \{0, 1, \ldots, 2^n - 1\} \rightarrow \{0, 1\}. \tag{60}$$

We require f be either *constant* or *balanced* as before. When f is constant, it takes a constant value 0 or 1 irrespetive of the input value x. When it is balanaced the value $f(x)$ for a half of $x \in S_n$ is 0 while it is 1 for the rest of x. Although there are functions which are neither constant nor balanced, we will not consider such cases here. Our task is to find an algorithm which tells if f is constant or balanced with the least possible number of evaluations of f.

It is clear that we need at least $2^{n-1} + 1$ steps, in the worst case with classical manipulations, to make sure if $f(x)$ is constant or balanced with 100 % confidence. It will be shown below that the number of steps reduces to a single step if we are allowed to use a quantum algorithm.

The algorithm is divided into the following steps:

(1) Prepare an $(n + 1)$-qubit register in the state $|\psi_0\rangle = |0\rangle^{\otimes n} \otimes |1\rangle$. First n qubits work as input qubits while the $(n + 1)$st qubit serves as a "scratch pad". Such qubits, which are neither input qubits nor output qubits, but work as a scratch pad to store temporary information are called ancillas or ancillary qubits.

(2) Apply the Walsh-Hadamard transforamtion to the register. Then we have the state

$$|\psi_1\rangle = U_{\mathrm{H}}^{\otimes n+1}|\psi_0\rangle = \frac{1}{\sqrt{2^n}}(|0\rangle + |1\rangle)^{\otimes n} \otimes \frac{1}{\sqrt{2}}(|0\rangle - |1\rangle)$$

$$= \frac{1}{\sqrt{2^n}} \sum_{x=0}^{2^n-1} |x\rangle \otimes \frac{1}{\sqrt{2}}(|0\rangle - |1\rangle). \tag{61}$$

(3) Apply the $f(x)$-controlled-NOT gate on the register, which flips the $(n+1)$st qubit if and only if $f(x) = 1$ for the input x. Therefore we need a U_f gate which evaluates $f(x)$ and acts on the register as $U_f|x\rangle|c\rangle = |x\rangle|c \oplus f(x)\rangle$, where $|c\rangle$ is the one-qubit state of the $(n + 1)$st qubit. Observe that $|c\rangle$ is flipped if and only if $f(x) = 1$ and left unchanged otherwise. We then obtain a state

$$|\psi_2\rangle = U_f|\psi_1\rangle = \frac{1}{\sqrt{2^n}} \sum_{x=0}^{2^n-1} |x\rangle \frac{1}{\sqrt{2}}(|f(x)\rangle - |\neg f(x)\rangle)$$

$$= \frac{1}{\sqrt{2^n}} \sum_x (-1)^{f(x)}|x\rangle \frac{1}{\sqrt{2}}(|0\rangle - |1\rangle). \tag{62}$$

Although the gate U_f is applied once for all, it is applied to *all* the n-qubit states $|x\rangle$ simultaneously.

(4) The Walsh-Hadamard transformation (46) is applied on the first n qubits next. We obtain

$$|\psi_3\rangle = (W_n \otimes I)|\psi_2\rangle = \frac{1}{\sqrt{2^n}} \sum_{x=0}^{2^n-1} (-1)^{f(x)} U_{\mathrm{H}}^{\otimes n}|x\rangle \frac{1}{\sqrt{2}}(|0\rangle - |1\rangle). \tag{63}$$

It is instructive to write the action of the one-qubit Hadamard gate as

$$U_{\mathrm{H}}|x\rangle = \frac{1}{\sqrt{2}}(|0\rangle + (-1)^x|1\rangle) = \frac{1}{\sqrt{2}} \sum_{y \in \{0,1\}} (-1)^{xy}|y\rangle,$$

where $x \in \{0, 1\}$, to find the resulting state. The action of the Walsh-

Hadamard transformation on $|x\rangle = |x_{n-1} \dots x_1 x_0\rangle$ yields

$$W_n|x\rangle = (U_H|x_{n-1}\rangle)(U_H|x_{n-2}\rangle) \dots (U_H|x_0\rangle)$$

$$= \frac{1}{\sqrt{2^n}} \sum_{y_{n-1}, y_{n-2}, \dots, y_0 \in \{0,1\}} (-1)^{x_{n-1}y_{n-1}+x_{n-2}y_{n-2}+\dots+x_0 y_0}$$

$$\times |y_{n-1}y_{n-2} \dots y_0\rangle = \frac{1}{\sqrt{2^n}} \sum_{y=0}^{2^n-1} (-1)^{x \cdot y}|y\rangle, \tag{64}$$

where $x \cdot y = x_{n-1}y_{n-1} \oplus x_{n-2}y_{n-2} \oplus \dots \oplus x_0 y_0$. Substituting this result into Eq. (63), we obtain

$$|\psi_3\rangle = \frac{1}{2^n}\left(\sum_{x,y=0}^{2^n-1} (-1)^{f(x)}(-1)^{x \cdot y}|y\rangle\right)\frac{1}{\sqrt{2}}(|0\rangle - |1\rangle). \tag{65}$$

(5) The first n qubits are measured. Suppose $f(x)$ is constant. Then $|\psi_3\rangle$ is put in the form

$$|\psi_3\rangle = \frac{1}{2^n}\sum_{x,y}(-1)^{x \cdot y}|y\rangle\frac{1}{\sqrt{2}}(|0\rangle - |1\rangle)$$

up to an overall phase. Let us consider the summation $\frac{1}{2^n}\sum_{x=0}^{2^n-1}(-1)^{x \cdot y}$ for a fixed $y \in S_n$. Clearly it vanishes since $x \cdot y$ is 0 for half of x and 1 for the other half of x unless $y = 0$. Therefore the summation yields δ_{y0}. Now the state reduces to $|\psi_3\rangle = |0\rangle^{\otimes n}\frac{1}{\sqrt{2}}(|0\rangle - |1\rangle)$ and the measurement outcome of the first n qubits is always $00 \dots 0$. Suppose $f(x)$ is balanced next. The probability amplitude of $|y = 0\rangle$ in $|\psi_3\rangle$ is proportional to $\sum_{x=0}^{2^n-1}(-1)^{f(x)}(-1)^{x \cdot y} = \sum_{x=0}^{2^n-1}(-1)^{f(x)} = 0$. Therefore the probability of obtaining measurement outcome $00 \dots 0$ for the first n qubits vanishes. In conclusion, the function f is constant if we obtain $00 \dots 0$ upon the meaurement of the first n qubits in the state $|\psi_3\rangle$ and it is balanced otherwise.

6. Decoherence

A quantum system is always in interaction with its environment. This interaction inevitably alter the state of the quantum system, which causes loss of information encoded in this system. The system under consideration is not a *closed* system when interaction with outside world is in action. We formulate the theory of *open* quantum system in this chapter by regarding the combined system of the quantum system and its environment as a closed system and subsequently trace out the environment degrees of freedom. Let ρ_S and ρ_E be the initial density matrices of the system and the environment, respectively. Even when the initial state is an uncorrelated state $\rho_S \otimes \rho_E$, the system-environment interaction entangles the total system so that the total state develops to an inseparable entangled state in general. Decoherence is a process in which environment causes various changes in the quantum system, which manifests itself as undesirable noise.

6.1. *Open quantum system*

Let us start our exposition with some mathematical background materials.[1,2,24]

We deal with general quantum states described by density matrices. We are interested in a general evolution of a quantum system, which is described by a powerful tool called a quantum operation. One of the simplest quantum operations is a unitary time evolution of a closed system. Let ρ_S be a density matrix of a closed system at $t = 0$ and let $U(t)$ be the time evolution operator. Then the corresponding quantum map $\mathcal{E}$ is defined as

$$\mathcal{E}(\rho_S) = U(t)\rho_S U(t)^\dagger. \tag{66}$$

One of our primary aims in this section is to generalize this map to cases of open quantum systems.

6.1.1. *Quantum operations and Kraus operators*

Suppose a system of interest is coupled with its environment. We must specify the details of the environment and the coupling between the system and the environment to study the effect of the environment on the behavior of the system. Let H_S, H_E and H_{SE} be the system Hamiltonian, the environment Hamiltonian and their interaction Hamiltonian, respectively. We assume the system-environment interaction is weak enough so that this separation into the system and its environment makes sense. To avoid confusion, we often call the system of interest the principal system. The total Hamiltonian H_T is then

$$H_T = H_S + H_E + H_{SE}. \tag{67}$$

Correspondingly, we denote the system Hilbert space and the environment Hilbert space as $\mathcal{H}_S$ and $\mathcal{H}_E$, respectively, and the total Hilbert space as $\mathcal{H}_T = \mathcal{H}_S \otimes \mathcal{H}_E$. The condition of weak system-environment interaction may be lifted in some cases. Let us consider a qubit propagating through a noisy quantum channel, for example. "Propagating" does not necessarily mean propagating in space. The qubit may be spatially fixed and subject to time-dependent noise. When the noise is localized in space and time, the input and the output qubit states belong to a well defined Hilbert space $\mathcal{H}_S$ and the above separation of the Hamiltonian is perfectly acceptable even for strongly interacting cases. We consider, in the following, how the principal system state ρ_S at $t = 0$ evolves in time in the presence of its environment. A map which describes a general change of the state from ρ_S to $\mathcal{E}(\rho_S)$ is called a quantum operation. We have already noted that the unitary time evolution is an example of a quantum operation. Other quantum operations include state change associated with measurement and state change due to noise. The latter quantum map is our primary interest in this chapter.

The state of the total system is described by a density matrix ρ. Suppose ρ is uncorrelated initally at time $t = 0$,

$$\rho(0) = \rho_S \otimes \rho_E, \tag{68}$$

where ρ_S (ρ_E) is the initial density matrix of the principal system (environment). The total system is assumed to be closed and to evolve with a unitary matrix $U(t)$ as

$$\rho(t) = U(t)(\rho_S \otimes \rho_E)U(t)^\dagger. \tag{69}$$

Note that the resulting state is not a tensor product state in general. We are interested in extracting information on the state of the principal system at some later time $t > 0$.

Even under these circumstances, however, we may still define the system density matrix $\rho_S(t)$ by taking partial trace of $\rho(t)$ over the environment Hilbert space as

$$\rho_S(t) = \text{tr}_E[U(t)(\rho_S \otimes \rho_E)U(t)^\dagger]. \tag{70}$$

We may forget about the environment by taking a trace over $\mathcal{H}_E$. This is an example of a quantum operation, $\mathcal{E}(\rho_S) = \rho_S(t)$. Let $\{|e_j\rangle\}$ be a basis of the system Hilbert space while $\{|\varepsilon_a\rangle\}$ be that of the environment Hilbert space. We may take the basis of $\mathcal{H}_T$ to be $\{|e_j\rangle \otimes |\varepsilon_a\rangle\}$. The initial density matrices may be written as $\rho_S = \sum_j p_j |e_j\rangle\langle e_j|$, $\rho_E = \sum_a r_a |\varepsilon_a\rangle\langle\varepsilon_a|$.

Action of the time evolution operator on a basis vector of $\mathcal{H}_T$ is explicitly written as

$$U(t)|e_j, \varepsilon_a\rangle = \sum_{k,b} U_{kb;ja}|e_k, \varepsilon_b\rangle, \tag{71}$$

where $|e_j, \varepsilon_a\rangle = |e_j\rangle \otimes |\varepsilon_a\rangle$ for example. Using this expression, the density matrix $\rho(t)$ is written as

$$U(t)(\rho_S \otimes \rho_E)U(t)^\dagger = \sum_{j,a} p_j r_a U(t)|e_j, \varepsilon_a\rangle\langle e_j, \varepsilon_a|U(t)^\dagger$$

$$= \sum_{j,a,k,b,l,c} p_j r_a U_{kb;ja}|e_k, \varepsilon_b\rangle\langle e_l, \varepsilon_c|U^*_{lc;ja}. \quad (72)$$

The partial trace over $\mathcal{H}_E$ is carried out to yield

$$\rho_S(t) = \mathrm{tr}_E[U(t)(\rho_S \otimes \rho_E)U(t)^\dagger] = \sum_{j,a,k,b,l} p_j r_a U_{kb;ja}|e_k\rangle\langle e_l|U^*_{lb;ja}$$

$$= \sum_{j,a,b} p_j \left(\sum_k \sqrt{r_a}U_{kb;ja}|e_k\rangle\right)\left(\sum_l \sqrt{r_a}\langle e_l|U^*_{lb;ja}\right). \quad (73)$$

To write down the quantum operation in a closed form, we assume the initial environment state is a pure state, which we take, without loss of generality, $\rho_E = |\varepsilon_0\rangle\langle\varepsilon_0|$. Even when ρ_E is a mixed state, we may always complement $\mathcal{H}_E$ with a fictitious Hilbert space to "purify" ρ_E, see § 2.7. With this assumption, $\rho_S(t)$ is written as

$$\rho_S(t) = \mathrm{tr}_E[U(t)(\rho_S \otimes |\varepsilon_0\rangle\langle\varepsilon_0|)U(t)^\dagger]$$

$$= \sum_a (I \otimes \langle\varepsilon_a|)U(t)(\rho_S \otimes |\varepsilon_0\rangle\langle\varepsilon_0|)|U(t)^\dagger(I \otimes |\varepsilon_a\rangle)$$

$$= \sum_a (I \otimes \langle\varepsilon_a|)U(t)(I \otimes |\varepsilon_0\rangle)\rho_S(I \otimes \langle\varepsilon_0|)U(t)^\dagger(I \otimes |\varepsilon_a\rangle).$$

We will drop $I\otimes$ from $I \otimes \langle\varepsilon_a|$ hereafter, whenever it does not cause confusion. Let us define the Kraus operator $E_a(t) : \mathcal{H}_S \to \mathcal{H}_S$ by

$$E_a(t) = \langle\varepsilon_a|U(t)|\varepsilon_0\rangle. \quad (74)$$

Then we may write

$$\mathcal{E}(\rho_S) = \rho_S(t) = \sum_a E_a(t)\rho_S E_a(t)^\dagger. \quad (75)$$

This is called the operator-sum representation (OSR) of a quantum operation $\mathcal{E}$. Note that $\{E_a\}$ satisfies the completeness relation

$$\left[\sum_a E_a(t)^\dagger E_a(t)\right]_{kl} = \left[\sum_a \langle\varepsilon_0|U(t)^\dagger|\varepsilon_a\rangle\langle\varepsilon_a|U(t)|\varepsilon_0\rangle\right]_{kl} = \delta_{kl}, \quad (76)$$

where I is the unit matrix in $\mathcal{H}_S$. This is equivalent with the trace-preserving property of $\mathcal{E}$ as $1 = \mathrm{tr}_S\rho_S(t) = \mathrm{tr}_S(\mathcal{E}(\rho_S)) = \mathrm{tr}_S\left(\sum_a E_a^\dagger E_a\rho_S\right)$ for any $\rho_S \in \mathcal{S}(\mathcal{H}_S)$. Completeness relation and trace-preserving property are satisfied since our total system is a closed system. A general quantum map does not necessarily satisfy these properties.[25]

At this stage, it turns out to be useful to relax the condition that $U(t)$ be a time evolution operator. Instead, we assume U be any operator including an arbitrary unitary gate. Let us consider a two-qubit system on which the CNOT gate acts. Suppose the principal system is the control qubit while the environment is the target qubit. Then we find

$$E_0 = (I \otimes \langle 0|)U_{\text{CNOT}}(I \otimes |0\rangle) = P_0, \; E_1 = (I \otimes \langle 1|)U_{\text{CNOT}}(I \otimes |0\rangle) = P_1,$$

where $P_i = |i\rangle\langle i|$, and consequently

$$\mathcal{E}(\rho_S) = P_0\rho_S P_0 + P_1\rho_S P_1 = \rho_{00}P_0 + \rho_{11}P_1 = \begin{pmatrix} \rho_{00} & 0 \\ 0 & \rho_{11} \end{pmatrix}, \quad (77)$$

where $\rho_S = \begin{pmatrix} \rho_{00} & \rho_{01} \\ \rho_{10} & \rho_{11} \end{pmatrix}$. Unitarity condition may be relaxed when measurements are included as quantum operations, for example.

Tracing out the extra degrees of freedom makes it impossible to invert a quantum operation. Given an initial principal system state ρ_S, there are infinitely many U that yield the same $\mathcal{E}(\rho_S)$. Therefore even though it is possible to compose two quantum operations, the set of quantum operations is not a group but merely a semigroup. [b]

6.1.2. Operator-sum representation and noisy quantum channel

Operator-sum representation (OSR) introduced in the previous subsection seems to be rather abstract. Here we give an interpretation of OSR as a noisy quantum channel. Suppose we have a set of unitary matrices $\{U_a\}$ and a set of non-negative real numbers $\{p_a\}$ such that $\sum_a p_a = 1$. By choosing U_a randomly with probability p_a and applying it to ρ_S, we define the expectation value of the resulting density matrix as

$$\mathcal{M}(\rho_S) = \sum_a p_a U_a \rho_S U_a^\dagger, \quad (78)$$

which we call a mixing process.[26] This occurs when a flying qubit is sent through a noisy quantum channel which transforms the density matrix by U_a with probability p_a, for example. Note that no enviroment has been introduced in the above definition, and hence no partial trace is involved.

---

[b] A set S is called a semigroup if S is closed under a product satisfying associativity $(ab)c = a(bc)$. If S has a unit element e, such that $ea = ae = a, \forall a \in S$, it is called a monoid.

Now the correspondence between $\mathcal{E}(\rho_S)$ and $\mathcal{M}(\rho_S)$ should be clear. Let us define $E_a \equiv \sqrt{p_a}U_a$. Then Eq. (78) is rewritten as

$$\mathcal{M}(\rho_S) = \sum_a E_a \rho_S E_a^\dagger \tag{79}$$

and the equivalence has been shown. Operators E_a are identified with the Kraus operators. The system transforms, under the action of U_a, as

$$\rho_S \to E_a \rho_S E_a^\dagger / \mathrm{tr}\left(E_a \rho_S E_a^\dagger\right). \tag{80}$$

Conversely, given a noisy quantum channel $\{U_a, p_a\}$ we may introduce an "environment" with the Hilbert space $\mathcal{H}_E$ as follows. Let $\mathcal{H}_E = \mathrm{Span}(|\varepsilon_a\rangle)$ be a Hilbert space with the dimension equal to the number of the unitary matrices $\{U_a\}$, where $\{|\varepsilon_a\rangle\}$ is an orthonormal basis. Define formally the environment density matrix $\rho_E = \sum_a p_a |\varepsilon_a\rangle\langle\varepsilon_a|$ and

$$U \equiv \sum_a U_a \otimes |\varepsilon_a\rangle\langle\varepsilon_a| \tag{81}$$

which acts on $\mathcal{H}_S \otimes \mathcal{H}_E$. It is easily verified from the orthonormality of $\{|\varepsilon_a\rangle\}$ that U is indeed a unitary matrix. Partial trace over $\mathcal{H}_E$ then yields

$$\begin{aligned}
\mathcal{E}(\rho_S) &= \mathrm{tr}_E[U(\rho_S \otimes \rho_E)U^\dagger] \\
&= \sum_a (I \otimes \langle\varepsilon_a|)\left(\sum_b U_b \otimes |\varepsilon_b\rangle\langle\varepsilon_b|\right)\left(\rho_S \otimes \sum_c p_c |\varepsilon_c\rangle\langle\varepsilon_c|\right) \\
&\quad \times \left(\sum_d U_d \otimes |\varepsilon_d\rangle\langle\varepsilon_d|\right)(I \otimes |\varepsilon_a\rangle) \\
&= \sum_a p_a U_a \rho_S U_a^\dagger = \mathcal{M}(\rho_S)
\end{aligned} \tag{82}$$

showing that the mixing process is also decribed by a quantum operation with a fictitious environment.

6.1.3. Completely positive maps

All linear operators we have encountered so far map vectors to vectors. A quantum operation maps a density matrix to another density matrix linearly.[c] A linear operator of this kind is called a superoperator. Let Λ be a superoperator acting on the system density matrices, $\Lambda : \mathcal{S}(\mathcal{H}_S) \to$

---

[c]Of course, the space of density matrices is not a linear vector space. What is meant hear is a linear operator, acting on the vector space of Hermitian matrices, also acts on the space of density matrices and it maps a density matrix to another density matrix.

$\mathcal{S}(\mathcal{H}_S)$. The operator Λ is easily extended to an operator acting on $\mathcal{H}_T$ by $\Lambda_T = \Lambda \otimes I_E$, which acts on $\mathcal{S}(\mathcal{H}_S \otimes \mathcal{H}_E)$. Note, however, that Λ_T is not necessarily a map $\mathcal{S}(\mathcal{H}_T) \to \mathcal{S}(\mathcal{H}_T)$. It may happen that $\Lambda_T(\rho)$ is not a density matrix any more. We have already encountered this situation when we have introduced partial transpose operation in § 2.7. Let $\mathcal{H}_T = \mathcal{H}_1 \otimes \mathcal{H}_2$ be a two-qubit Hilbert space, where $\mathcal{H}_k$ is the kth qubit Hilbert space. It is clear that the transpose operation $\Lambda_t : \rho_1 \to \rho_1^t$ on a single-qubit state ρ_1 preserves the density matrix properties. For a two-qubit density matrix ρ_{12}, however, this is not always the case. In fact, we have seen that $\Lambda_t \otimes I : \rho_{12} \to \rho_{12}^{\mathrm{pt}}$ defined by Eq. (22) maps a density matrix to a matrix which is not a density matrix when ρ_{12} is inseparable.

A map Λ which maps a positive operator acting on $\mathcal{H}_S$ to another positive operator on $\mathcal{H}_S$ is said to be positive. Moreover, it is called a completely positive map (CP map), if its extension $\Lambda_T = \Lambda \otimes I_n$ remains a positive operator for an arbitrary $n \in \mathbb{N}$.

Theorem 6.1. *A linear map Λ is CP if and only if there exists a set of operators $\{E_a\}$ such that $\Lambda(\rho_S)$ can be written as*

$$\Lambda(\rho_S) = \sum_a E_a \rho_S E_a^\dagger. \tag{83}$$

We require not only that Λ be CP but also $\Lambda(\rho)$ be a density matrix:

$$\mathrm{tr}\,\Lambda(\rho_S) = \mathrm{tr}\left(\sum_a E_a \rho E_a^\dagger\right) = \mathrm{tr}\left(\sum_a E_a^\dagger E_a \rho\right) = 1. \tag{84}$$

This condition is satisfied for any ρ if and only if

$$\sum_a E_a^\dagger E_a = I_S. \tag{85}$$

Therefore, any quantum operation obtained by tracing out the environment degrees of freedom is CP and preserves trace.

6.2. *Measurements as quantum operations*

We have already seen that a unitary evoluation $\rho_S \to U\rho_S U^\dagger$ and a mixing process $\rho_S \to \sum_i p_i U_i \rho_S U_i^\dagger$ are quantum operations. We will see further examples of quantum operations in this section and the next. This section deals with measurements as quantum operations.

6.2.1. *Projective measurements*

Suppose we measure an observable $A = \sum_i \lambda_i P_i$, where $P_i = |\lambda_i\rangle\langle\lambda_i|$ is the projection operator corresponding to the eigenvector $|\lambda_i\rangle$. We have seen in Chapter 2 that the probability of observing λ_i upon a measurement of A in a state ρ is

$$p(i) = \langle\lambda_i|\rho|\lambda_i\rangle = \mathrm{tr}\,(P_i\rho) \tag{86}$$

and the state changes as $\rho \to P_i\rho P_i/p(i)$. This process happens with a probability $p(i)$. Thus we may regard the measurement process as a quantum operation

$$\rho_S \to \sum_i p(i)\frac{P_i\rho_S P_i}{p(i)} = \sum_i P_i\rho_S P_i, \tag{87}$$

where the set $\{P_i\}$ satisifes the completeness relation $\sum_i P_i P_i^\dagger = I$.

The projective measurement is a special case of a quantum operation in which the Kraus operators are $E_i = P_i$.

6.2.2. *POVM*

We have been concerned with projective measurements so far. However, it should be noted that they are not unique type of measurements. Here we will deal with the most general framework of measurement and show that it is a quantum operation.

Suppose a system and an environment, prepared initially in a product state $|\psi\rangle|e_0\rangle$, are acted by a unitary operator U, which applies an operator M_i on the system and, at the same time, put the environment to $|e_i\rangle$ for various i. It is written explicitly as

$$|\Psi\rangle = U|\psi\rangle|e_0\rangle = \sum_i M_i|\psi\rangle|e_i\rangle. \tag{88}$$

The system and its environment are correlated in this way. This state must satisfy the normalization condition since U is unitary; $\langle\psi|\langle e_0|U^\dagger U|\psi\rangle|e_0\rangle = \sum_{i,j}\langle\psi|\langle e_i|M_i^\dagger M_j \otimes I|\psi\rangle|e_j\rangle = \langle\psi|\sum_i M_i^\dagger M_i|\psi\rangle = 1$. Since $|\psi\rangle$ is arbitrary, we must have

$$\sum_i M_i^\dagger M_i = I_S, \tag{89}$$

where I_S is the unit matrix acting on the system Hilbert space $\mathcal{H}_S$. Operators $\{M_i^\dagger M_i\}$ are said to form a POVM (positive operator-valued measure).

Suppose we measure the environment with a measurement operator

$$O = I_S \otimes \sum_i \lambda_i |e_i\rangle\langle e_i| = \sum_i \lambda_i \left(I_S \otimes |e_i\rangle\langle e_i|\right).$$

We obtain a measurement outcome λ_k with a probability

$$\begin{aligned}
p(k) &= \langle\Psi|(I_S \otimes |e_k\rangle\langle e_k|)|\Psi\rangle \\
&= \sum_{i,j}\langle\psi|\langle e_i|M_i^\dagger(I_S \otimes |e_k\rangle\langle e_k|)M_j|\psi\rangle|e_j\rangle = \langle\psi|M_k^\dagger M_k|\psi\rangle, \quad (90)
\end{aligned}$$

where $|\Psi\rangle = U|\psi\rangle|e_0\rangle$. The combined system immediately after the measurement is

$$\begin{aligned}
\frac{1}{\sqrt{p(k)}}(I_S \otimes |e_k\rangle\langle e_k|)U|\psi\rangle|e_0\rangle &= \frac{1}{\sqrt{p(k)}}(I_S \otimes |e_k\rangle\langle e_k|)\sum_i M_i|\psi\rangle|e_i\rangle \\
&= \frac{1}{\sqrt{p(k)}}M_k|\psi\rangle|e_k\rangle. \quad (91)
\end{aligned}$$

Let $\rho_S = \sum_i p_i|\psi_i\rangle\langle\psi_i|$ be an arbitrary density matrix of the principal system. It follows from the above observation for a pure state $|\psi\rangle\langle\psi|$ that the reduced density matrix immediately after the measurement is

$$\sum_k p(k)\frac{M_k\rho_S M_k^\dagger}{p(k)} = \sum_k M_k\rho_S M_k^\dagger. \quad (92)$$

This shows that POVM measurement is a quantum operation in which the Kraus operators are given by the generalized measurement operators $\{M_i\}$. The projective measurement is a special class of POVM, in which $\{M_i\}$ are the projective operators.

6.3. *Examples*

Now we examine several important examples which have relevance in quantum information theory. Decoherence appears as an error in quantum information processing. The next chapter is devoted to strategies to fight against errors introduced in this section.

6.3.1. *Bit-flip channel*

Consider a closed two-qubit system with a Hilbert space $\mathbb{C}^2 \otimes \mathbb{C}^2$. We call the first qubit the "(principal) system" while the second qubit the "environment". A bit-flip channel is defined by a quantum operation

$$\mathcal{E}(\rho_S) = (1-p)\rho_S + p\sigma_x\rho_S\sigma_x, \quad 0 \leq p \leq 1. \quad (93)$$

The input ρ_S is bit-flipped with a probability p while it remains in its input state with a probability $1 - p$. The Kraus operators are read off as

$$E_0 = \sqrt{1-p}I, \ E_1 = \sqrt{p}\sigma_x. \tag{94}$$

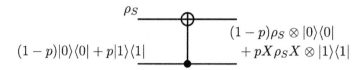

Fig. 4. Quantum circuit modelling a bit-flip channel. The gate is the inverted CNOT gate $I \otimes |0\rangle\langle 0| + \sigma_x \otimes |1\rangle\langle 1|$.

The circuit depicted in Fig. 4 models the bit-flip channel provided that the second qubit is in a mixed state $(1 - p)|0\rangle\langle 0| + p|1\rangle\langle 1|$. The circuit is nothing but the inverted CNOT gate $V = I \otimes |0\rangle\langle 0| + \sigma_x \otimes |1\rangle\langle 1|$. The output of this circuit is

$$\begin{aligned}
V\left(\rho_S \otimes [(1-p)|0\rangle\langle 0| + p|1\rangle\langle 1|]\right) V^\dagger \\
= (1-p)\rho_S \otimes |0\rangle\langle 0| + p\sigma_x\rho_S\sigma_x|1\rangle\langle 1|,
\end{aligned} \tag{95}$$

from which we obtain

$$\mathcal{E}(\rho_S) = (1-p)\rho_S + p\sigma_x\rho_S\sigma_x \tag{96}$$

after tracing over the environment Hilbert space.

The choice of the second qubit input state is far from unique and so is the choice of the circuit. Suppose the initial state of the environment is a pure state $|\psi_E\rangle = \sqrt{1-p}|0\rangle + \sqrt{p}|1\rangle$, for example. Then the output of the circuit in Fig. 4 is

$$\mathcal{E}(\rho_S) = \text{tr}_E[V\rho_S \otimes |\psi_E\rangle\langle\psi_E|V^\dagger] = (1-p)\rho_S + p\sigma_x\rho_S\sigma_x, \tag{97}$$

producing the same result as before.

Let us see what transformation this quantum operation brings about in ρ_S. We parametrize ρ_S using the Bloch vector as

$$\rho_S = \frac{1}{2}\left(I + \sum_{k=x,y,z} c_k\sigma_k\right), \quad (c_k \in \mathbb{R}) \tag{98}$$

where $\sum_k c_k^2 \le 1$. We obtain

$$\mathcal{E}(\rho_S) = (1-p)\rho_S + p\sigma_x\rho_S\sigma_x$$
$$= \frac{1-p}{2}(I + c_x\sigma_x + c_y\sigma_y + c_z\sigma_z) + \frac{p}{2}(I + c_x\sigma_x - c_y\sigma_y - c_z\sigma_z)$$
$$= \frac{1}{2}\begin{pmatrix} 1+(1-2p)c_z & c_x - i(1-2p)c_y \\ c_x + i(1-2p)c_y & 1 - (1-2p)c_z \end{pmatrix}. \tag{99}$$

Observe that the radius of the Bloch sphere is reduced along the y- and the z-axes so that the radius in these directions is $|1-2p|$. Equation (99) shows that the quantum operation has produced a mixture of the Bloch vector states (c_x, c_y, c_z) and $(c_x, -c_y, -c_z)$ with weights $1-p$ and p respectively. Figure 5 (a) shows the Bloch sphere which represents the input qubit states. The Bloch sphere shrinks along the y- and z-axes, which results in the ellipsoid shown in Fig. 5 (b).

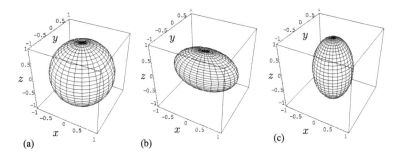

Fig. 5. Bloch sphere of the input state ρ_S (a) and output states of (b) bit-flip channel and (c) phase-flip channel. The probability $p = 0.2$ is common to both channels.

6.3.2. *Phase-flip channel*

Consider again a closed two-qubit system with the "(principal) system" and its "environment".

The phase-flip channel is defined by a quantum operation

$$\mathcal{E}(\rho_S) = (1-p)\rho_S + p\sigma_z\rho_S\sigma_z, \ 0 \le p \le 1. \tag{100}$$

The input ρ_S is phase-flipped ($|0\rangle \mapsto |0\rangle$ and $|1\rangle \mapsto -|1\rangle$) with a probability p while it remains in its input state with a probability $1-p$. The corresponding Kraus operators are

$$E_0 = \sqrt{1-p}I, \ E_1 = \sqrt{p}\sigma_z. \tag{101}$$

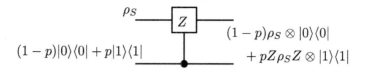

Fig. 6. Quantum circuit modelling a phase-flip channel. The gate is the inverted controlled-σ_z gate.

A quantum circuit which models the phase-flip channel is shown in Fig. 6. Let ρ_S be the first qubit input state while $(1-p)|0\rangle\langle 0| + p|1\rangle\langle 1|$ be the second qubit input state. The circuit is the inverted controlled-σ_z gate

$$V = I \otimes |0\rangle\langle 0| + \sigma_z \otimes |1\rangle\langle 1|.$$

The output of this circuit is

$$V\left(\rho_S \otimes [(1-p)|0\rangle\langle 0| + p|1\rangle\langle 1|]\right)V^\dagger$$
$$= (1-p)\rho_S \otimes |0\rangle\langle 0| + p\sigma_z\rho_S\sigma_z \otimes |1\rangle\langle 1|, \qquad (102)$$

from which we obtain

$$\mathcal{E}(\rho_S) = (1-p)\rho_S + p\sigma_z\rho_S\sigma_z. \qquad (103)$$

The second qubit input state may be a pure state

$$|\psi_E\rangle = \sqrt{1-p}|0\rangle + \sqrt{p}|1\rangle, \qquad (104)$$

for example. Then we find

$$\mathcal{E}(\rho_S) = \mathrm{tr}_E[V\rho_S \otimes |\psi_E\rangle\langle\psi_E|V^\dagger] = E_0\rho_S E_0^\dagger + E_1\rho_S E_1^\dagger, \qquad (105)$$

where the Kraus operators are

$$E_0 = \langle 0|V|\psi_E\rangle = \sqrt{1-p}I, \; E_1 = \langle 1|V|\psi_E\rangle = \sqrt{p}\sigma_z. \qquad (106)$$

Let us work out the transformation this quantum operation brings about to ρ_S. We parametrize ρ_S using the Bloch vector as before. We obtain

$$\mathcal{E}(\rho_S) = (1-p)\rho_S + p\sigma_z\rho_S\sigma_z$$
$$= \frac{1-p}{2}(I + c_x\sigma_x + c_y\sigma_y + c_z\sigma_z) + \frac{p}{2}(I - c_x\sigma_x - c_y\sigma_y + c_z\sigma_z)$$
$$= \frac{1}{2}\begin{pmatrix} 1+c_z & (1-2p)(-c_x - ic_y) \\ (1-2p)(c_x + ic_y) & 1-c_z \end{pmatrix}. \qquad (107)$$

Observe that the off-diagonal components decay while the diagonal components remain the same. Equation (107) shows that the quantum operation has produced a mixture of the Bloch vector states (c_x, c_y, c_z) and

$(-c_x, -c_y, c_z)$ with weights $1 - p$ and p respectively. The initial state has a definite phase $\phi = \tan^{-1}(c_y/c_x)$ in the off-diagonal components. The phase after the quantum operaition is applied is a mixture of states with ϕ and $\phi + \pi$. This process is called the phase relaxation process, or the T_2 process in the context of NMR. The radius of the Bloch sphere is reduced along the x- and the y-axes as $1 \to |1 - 2p|$. Figure 5 (c) shows the effect of the phase-flip channel on the Bloch sphere for $p = 0.2$.

Other examples will be found in [1,2].

7. Quantum Error Correcting Codes

7.1. *Introduction*

It has been shown in the previous chapter that interactions between a quantum system with environment cause undesirable changes in the state of the quantum system. In the case of qubits, they appear as bit-flip and phase-flip errors, for example. To reduce such errors, we must implement some sort of error correcting mechanism in the algorithm.

Before we introduce quantum error correcting codes, we have a brief look at the simplest version of error correcting code in classical bits. Suppose we transmit a serise of 0's and 1's through a noisy classical channel. Each bit is assumed to flip independently with a probability p. Thus a bit 0 sent through the channel will be received as 0 with probability $1 - p$ and as 1 with probability p. To reduce channel errors, we may invoke to majority vote. Namely, we encode logical 0 by 000 and 1 by 111, for example. When 000 is sent through this channel, it will be received as 000 with probability $(1-p)^3$, as $100, 010$ or 001 with probability $3p(1-p)^2$, as $011, 101$ or 110 with probability $3p^2(1 - p)$ and finally as 111 with probability p^3. By taking the majority vote, we correctly reproduce the desired result 0 with probability $p_0 = (1 - p)^3 + 3p(1 - p)^2 = (1 - p)^2(1 + 2p)$ while fails with probability $p_1 = 3p^2(1 - p) + p^3 = (3 - 2p)p^2$. We obtain $p_0 \gg p_1$ for sufficiently small $p \geq 0$. In fact, we find $p_0 = 0.972$ and $p_1 = 0.028$ for $p = 0.1$. The success probability p_0 increases as p approaches to 0, or alternatively, if we use more bits to encode 0 or 1.

This method cannot be applicable to qubits, however, due to no-cloning theorem. We have to somehow think out the way to overcome this theorem.

7.2. *Three-qubit bit-flip code: the simplest example*

It is instructive to introduce a simple example of quantum error correcting codes (QECC). We closely follow Steane[30] here.

7.2.1. *Bit-flip QECC*

Suppose Alice wants to send a qubit or a series of qubits to Bob through a noisy quantum channel. Let $|\psi\rangle = a|0\rangle + b|1\rangle$ be the state she wants to send. If she is to transmit a serise of qubits, she sends them one by one and the following argument applies to each of the qubits. Let p be the probability with which a qubit is flipped and we assume there are no other types of errors in the channel. In other words, the operator X is applied to the qubit with probability p and consequently the state is mapped to

$$|\psi\rangle \to |\psi'\rangle = X|\psi\rangle = a|1\rangle + b|0\rangle. \tag{108}$$

We have already seen in the previous section that this channel is described by a quantum operation (93).

7.2.2. *Encoding*

To reduce the error probability, we want to mimic somehow the classical counterpart without using a clone machine. Let us recall that the action of a CNOT gate is CNOT : $|j0\rangle \to |jj\rangle$, $j \in \{0, 1\}$ and therefore it duplicates the control bit $j \in \{0, 1\}$ when the target bit is initially set to $|0\rangle$. We use this fact to *triplicate* the basis vectors as

$$|\psi\rangle|00\rangle = (a|0\rangle + b|1\rangle)|00\rangle \to |\psi\rangle_E = a|000\rangle + b|111\rangle, \tag{109}$$

where $|\psi\rangle_E$ denotes the encoded state. The state $|\psi\rangle_E$ is called the logical qubit while each constituent qubit is called the physical qubit. We borrow terminologies from classical error correcting code (ECC) and call the set

$$C = \{a|000\rangle + b|111\rangle | a, b \in \mathbb{C}, |a|^2 + |b|^2 = 1\} \tag{110}$$

the code and each member of C a codeword. It is important to note that the state $|\psi\rangle$ is not triplicated but only the basis vectors are triplicated. This redundancy makes it possible to detect errors in $|\psi\rangle_E$ and correct them as we see below.

A quantum circuit which implements the encoding (109) is easily found from our experience in CNOT gate. Let us consider the circuit shown in Fig. 7 (a) whose input state is $|\psi\rangle|00\rangle$. It is immediately found that the output of this circuit is $|\psi\rangle_E = a|000\rangle + b|111\rangle$ as promised.

7.2.3. *Transmission*

Now the state $|\psi\rangle_E$ is sent through a quantum channel which introduces bit-flip error with a rate p for each qubit independently. We assume p is

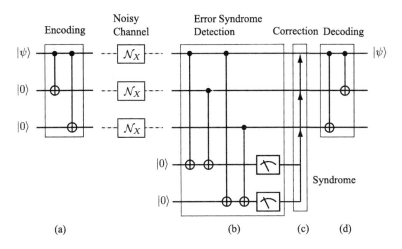

Fig. 7. Quantum circuits to (a) encode, (b) detect bit-flip error syndrome, (c) make correction to a relevant qubit and (d) decode. The gate $\mathcal{N}_X$ stands for the bit-flip noise.

sufficiently small so that not many errors occur during qubit transmission. The received state depends on in which physical qubit(s) the bit-flip error occurred. Table 1 lists possible received states and the probabilities with which these states are received.

Table 1. State Bob receives and the probability which this may happen.

State Bob receives	Probability
$a\|000\rangle + b\|111\rangle$	$(1-p)^3$
$a\|100\rangle + b\|011\rangle$	$p(1-p)^2$
$a\|010\rangle + b\|101\rangle$	$p(1-p)^2$
$a\|001\rangle + b\|110\rangle$	$p(1-p)^2$
$a\|110\rangle + b\|001\rangle$	$p^2(1-p)$
$a\|101\rangle + b\|010\rangle$	$p^2(1-p)$
$a\|011\rangle + b\|100\rangle$	$p^2(1-p)$
$a\|111\rangle + b\|000\rangle$	p^3

7.2.4. Error syndrome dectection and correction

Now Bob has to extract from the received state which error occurred during qubits transmission. For this purpose, Bob prepares two ancillary qubits in the state $|00\rangle$ as depicted in Fig. 7 (b) and apply four CNOT operations whose control bits are the encoded qubits while the target qubits are Bob's two ancillary qubits. Let $|x_1x_2x_3\rangle$ be a basis vectors Bob has received and let A (B) be the output state of the first (second) ancilla qubit. It is seen from Fig. 7 (b) that $A = x_1 \oplus x_2$ and $B = x_1 \oplus x_3$. Let $a|100\rangle + b|011\rangle$ be the received logical qubit for example. Note that the first qubit state in both of the basis vectors is different from the second and the third qubit states. These difference are detected by the pairs of CNOT gates in Fig. 7 (b). The error extracting sequence transforms the anicillary qubits as

$$(a|100\rangle + b|011\rangle)|00\rangle \to a|10011\rangle + b|01111\rangle = (a|100\rangle + b|011\rangle)|11\rangle.$$

Both of the ancillary qubits are flipped since $x_1 \oplus x_2 = x_1 \oplus x_3 = 1$ for both $|100\rangle$ and $|011\rangle$. It is important to realize that (i) the syndrome is independent of a and b and (ii) the received state $a|100\rangle + b|011\rangle$ remains intact; we have detected an error without measuring the received state! These features are common to all QECC.

We list the result of other cases in Table 2. Note that among eight

Table 2.   States after error extraction is made and the probabilities with which these states are produced.

State after error syndrome extraction	Probability
$(a|000\rangle + b|111\rangle)|00\rangle$	$(1-p)^3$
$(a|100\rangle + b|011\rangle)|11\rangle$	$p(1-p)^2$
$(a|010\rangle + b|101\rangle)|10\rangle$	$p(1-p)^2$
$(a|001\rangle + b|110\rangle)|01\rangle$	$p(1-p)^2$
$(a|110\rangle + b|001\rangle)|01\rangle$	$p^2(1-p)$
$(a|101\rangle + b|010\rangle)|10\rangle$	$p^2(1-p)$
$(a|011\rangle + b|100\rangle)|11\rangle$	$p^2(1-p)$
$(a|111\rangle + b|000\rangle)|00\rangle$	p^3

possible states, there are exactly two states with the same ancilla state. Does it mean this error extraction scheme does not work? Now let us compare the probabilities associated with the same ancillary state. When the ancillary state is $|10\rangle$, for example, there are two possible reseived states $a|010\rangle + b|101\rangle$ and $a|101\rangle + b|010\rangle$. Note that the former is received with probability $p(1-p)^2$ while that latter with $p^2(1-p)$. Therefore the latter probability is negligible compared to the former for sufficiently small p.

It is instructive to visualize what errors do to the encoded basis vectors. Consider a cube with the unit length. The vertices of the cube have coordinates (i, j, k) where $i, j, k \in \{0, 1\}$. We assign a vector $|ijk\rangle$ to the vertex (i, j, k), under which the vectors $|000\rangle$ and $|111\rangle$ correspond to diagonally separated vertices. An action of X_i, the operator $X = \sigma_x$ acting on the ith qubit, sends these basis vectors to the nearest neighbor vetices, which differ from the correct basis vectors in the ith position. The intersection of the sets of vectors obtained by a single action of X_i on $|000\rangle$ and $|111\rangle$ is an empty set. Therefore an action of a single error operator X can be corrected with no ambiguity.

Now Bob measures his ancillary qubits and obtains two bits of classical information. The set of two bits is called the (error) syndrome and it tells Bob in which physical qubit the error occurred during transmission. Bob applies correcting procedure to the received state according to the error sydrome he has obtained. Ignoring extra error states with small probabilities, we immediately find that the following action must be taken:

error syndrome	correction to be made
00	identity operation (nothing is required)
01	apply σ_x to the third qubit
10	apply σ_x to the second qubit
11	apply σ_x to the first qubit

Suppose the syndrome is 01, for example. The state Bob received is likely to be $a|001\rangle + b|110\rangle$. Bob recovers the initial state Alice has sent by applying $I \otimes I \otimes \sigma_x$ on the received state:

$$(I \otimes I \otimes \sigma_x)(a|001\rangle + b|110\rangle) = a|000\rangle + b|111\rangle.$$

If Bob receives the state $a|110\rangle + b|001\rangle$, unfortunately, he will obtain

$$(I \otimes I \otimes \sigma_x)(a|110\rangle + b|001\rangle) = a|111\rangle + b|000\rangle.$$

In fact, for any error sydrome, Bob obtaines either $a|000\rangle+b|111\rangle$ or $a|111\rangle+b|000\rangle$. The latter case occurs if and only if more than one qubit are flipped, and hence it is less likely to happen for sufficiently small error rate p. The probability with which multiple error occurs is found from Table 1 as

$$P(\text{error}) = 3p^2(1 - p) + p^3 = 3p^2 - 2p^3. \tag{111}$$

This error rate is less than p if $p < 1/2$. In contrast, success probability has been enhanced from $1 - p$ to $1 - P(\text{error}) = 1 - 3p^2 + 2p^3$. Let $p = 0.1$, for example. Then the error rate is lowered to $P(\text{error}) = 0.028$, while the success probability is enhanced from 0.9 to 0.972.

7.2.5. *Decoding*

Now that Bob has corrected an error, what is left for him is to decode the encoded state. This is nothing but the inverse transformation of the encoding (109). It can be seen from Fig. 7 (d) that

$$\text{CNOT}_{12}\text{CNOT}_{13}(a|000\rangle + b|111\rangle) = a|000\rangle + b|100\rangle = (a|0\rangle + b|1\rangle)|00\rangle. \tag{112}$$

7.2.6. Miracle of entanglement

This example, albeit simple, contains almost all fundamental ingredients of QECC. We prepare some redundant qubits which somehow "triplicate" the original qubit state to be sent without violating no-cloning theorem. Then the encoded qubits are sent through a noisy channel, which causes a bit-flip in at most one of the qubits. The received state, which may be subject to an error, is then entangled with ancillary qubits, whose state reflects the error which occurred during the state transmission. This results in an entangled state

$$\sum_k |\text{A bit-flip error in the } k\text{th qubit}\rangle \otimes |\text{corresponding error syndrome}\rangle. \tag{113}$$

The wave function, upon the measurement of the ancillary qubits, collapses to a state with a bit-flip error corresponding to the observed error syndrome. In a sense, syndrome measurement singles out a particular error state which produces the observed syndrome.

Once syndrome is found, it is an easy task to transform the received state back to the original state. Note that everything is done without knowing what the origial state is.

7.2.7. Continuous rotations

We have considered noise X so far. Suppose noise in the channel is characterized by a contiuous paramter α as

$$U_\alpha = e^{i\alpha X} = \cos\alpha I + iX\sin\alpha, \tag{114}$$

which maps a state $|\psi\rangle$ to

$$U_\alpha|\psi\rangle = \cos\alpha|\psi\rangle + i\sin\alpha X|\psi\rangle. \tag{115}$$

Suppose U_α acts on the first qubit, for example. Bob then receives

$$(U_\alpha \otimes I \otimes I)(a|000\rangle + b|111\rangle)$$
$$= \cos\alpha(a|000\rangle + b|111\rangle) + i\sin\alpha(a|100\rangle + b|011\rangle).$$

The output of the error syndrome detection circuit, before the syndrome measurement is made, is an entangled state

$$\cos\alpha(a|000\rangle + b|111\rangle)|00\rangle + i\sin\alpha(a|100\rangle + b|111\rangle)|11\rangle, \qquad (116)$$

see Table 2. Measurement of the error syndrome yields either 00 or 11. In the former case the state collapses to $|\psi\rangle = a|000\rangle + b|111\rangle$ and this happens with a probability $\cos^2\alpha$. In the latter case, on the other hand, the received state collapses to $X|\psi\rangle = a|100\rangle + b|011\rangle$ and this happens with a probability $\sin^2\alpha$. Bob applies I (X) to the first qubit to correct the error when the syndrome readout is 00 (11).

It is clear that error U_α may act on the second or the third qubit. Continuous rotation U_α for any α may be corrected in this way. In general, linearity of a quantum circuit guarantees that any QECC, which corrects the bit-flip error X, corrects continuous error U_α.

8. DiVincenzo Criteria

We have learned so far that information may be encoded and processed in a quantum-mechanical way. This new discipline called quantum information processing (QIP) is expected to solve a certain class of problems that current digital computers cannot solve in a practical time scale. Although a small scale quantum information processor is already available commercially, physical realization of large scale quantum information processors is still beyond the scope of our currently available technology.

A quantum computer should have at least $10^2 \sim 10^3$ qubits to be able to execute algorithms that are more efficient than their classical counterparts. DiVincenzo proposed necessary conditions, so-called the *DiVincenzo criteria* that any physical system has to fulfill to be a candidate for a viable quantum computer.[38] In the next section, we outline these conditions as well as two additional criteria for networkability.

8.1. *DiVincenzo criteria*

In his influential article,[38] DiVincenzo proposed five criteria that any physical system must satisfy to be a viable quantum computer. We summarize the relevant parts of these criteria in this section.

(1) *A scalable physical system with well characterized qubits.*

To begin with, we need a quantum register made of many qubits to store information. Recall that a classical computer also requires memory to store information. The simplest way to realize a qubit physically is to use a two-level quantum system. For example, an electron, a spin $1/2$ nucleus or two mutually orthogonal polarization states (horizontal and vertical, for example) of a single photon can be a qubit. We may also employ a two-dimensional subspace, such as the ground state and the first excited state, of a multi-dimensional Hilbert space, such as atomic energy levels. In any case, the two states are identified as the basis vectors, $|0\rangle$ and $|1\rangle$, of the Hilbert space so that a general single qubit state takes the form $|\psi\rangle = \alpha|0\rangle + \beta|1\rangle$, $|\alpha|^2 + |\beta|^2 = 1$. A multi-qubit state is expanded in terms of the tensor products of these basis vectors. Each qubit must be separately addressable. Moreover it should be scalable up to a large number of qubits. The two-dimensional vector space of a qubit may be extended to three-dimensional (qutrit) or, more generally, d-dimensional (qudit).

A system may be made of several different kinds of qubits. Qubits in an ion trap quantum computer, for instance, may be defined as: (1) hyperfine/Zeeman sublevels in the electronic ground state of ions (2) a ground state and an excited state of a weakly allowed optical transition and (3) normal mode of ion oscillation. A similar scenario is also proposed for Josephson junction qubits, in which two flux qubits are coupled through a quantized LC circuit. Simultaneous usage of several types of qubits may be the most promising way to achieving a viable quantum computer.

(2) *The ability to initialize the state of the qubits to a simple fiducial state, such as $|00\ldots0\rangle$.*

Suppose you are not able to reset your (classical) computer. Then you will never trust the output of some computation even though processing is done correctly. Therefore initialization is an important part of both quantum and classical information processors.

In many realizations, initialization may be done simply by cooling to bring the system into its ground state. Let ΔE be the difference between energies of the first excited state and the ground state. The system is in the ground state with a good precision at low temperatures satisfying $k_B T \ll \Delta E$. Alternatively, we may use projective measurement to project the system onto a desired state. In some cases, we observe the system to be in an undesired state upon such measurement. Then we may transform the system to the desired fiducial state by applying appropriate gates.

For some realizations, such as liquid state NMR, however, it is impossible to cool the system down to extremely low temperatures. In those cases, we are forced to use a thermally populated state as an initial state. This seemingly difficult problem may be amended by several methods if some computational resources are sacrificed. We then obtain an "effective" pure state, so-called the pseudopure state, which works as an initial state for most purposes.

Continuous fresh supply of qubits in a specified state, such as $|0\rangle$, is also an important requirement for successful quantum error correction. as we have seen in Section 7.

(3) *Long decoherence times, much longer than the gate operation time.*

Hardware of a classical computer lasts long, for on the order of 10 years. Things are totally different for a quantum computer, which is fragile against external disturbance called decoherence, see Section 6.

Decoherence is probably the hardest obstacle to building a viable quantum computer. Decoherence means many aspects of quantum state degradation due to interactions of the system with the environment and sets the maximum time available for quantum computation. Decoherence time itself is not very important. What matters is the ratio "decoherence time/gate operation time". For some realizations, decoherence time may be as short as $\sim \mu$s. This is not necessarily a big problem provided that the gate operation time, determined by the Rabi oscillation period and the qubit-coupling strength, for example, is much shorter than the decoherence time. If the typical gate operation time is $\sim$ ps, say, the system may execute $10^{12-6} = 10^6$ gate operations before the quantum state decays. We quote the number $\sim 10^5$ of gates required to factor 21 into 3 and 7 by using Shor's algorithm.[40]

There are several ways to effectively prolong decoherence time. A closed-loop control method incorporates QECC, while an open-loop control method incorporates noiseless subsystem[41] and decoherence free subspace (DFS).[42]

(4) *A "universal" set of quantum gates.*

Suppose you have a classical computer with a big memory. Now you have to manipulate the data encoded in the memory by applying various logic gates. You must be able to apply arbitrary logic operations on the memory bits to carry out useful information processing. It is known that the NAND gate is universal, i.e., any logic gates may be implemented with NAND gates.

Let $H(\gamma(t))$ be the Hamiltonian of an n-qubit system under consideration, where $\gamma(t)$ collectively denotes the control parameters in the Hamiltonian. The time-development operator of the system is $U[\gamma(t)] = \mathcal{T} \exp\left[-\frac{i}{\hbar} \int^T H(\gamma(t)) dt\right] \in U(2^n)$, where $\mathcal{T}$ is the time-ordering operator. Our task is to find the set of control parameters $\gamma(t)$, which implements the desired gate U_{gate} as $U[\gamma(t)] = U_{\text{gate}}$. Although this "inverse problem" seems to be difficult to solve, a theorem by Barenco et al. guarantees that any $U(2^n)$ gate may be decomposed into single-qubit gates $\in U(2)$ and CNOT gates.[16] Therefore it suffices to find the control sequences to implement $U(2)$ gates and a CNOT gate to construct an arbitrary gate. Naturally, implementation of a CNOT gate in any realization is considered to be a milestone in this respect. Note, however, that any two-qubit gates, which are neither a tensor product of two one-qubit gates nor a SWAP gate, work as a component of a universal set of gates.[43]

(5) *A qubit-specific measurement capability.*

The result of classical computation must be displayed on a screen or printed on a sheet of paper to readout the result. Although the readout process in a classical computer is regarded as too trivial a part of computation, it is a vital part in quantum computing.

The state at the end of an execution of quantum algorithm must be measured to extract the result of the computation. The measurement process depends heavily on the physical system under consideration. For most realizations, projective measurements are the primary method to extract the outcome of a computation. In liquid state NMR, in contrast, a projective measurement is impossible, and we have to resort to ensemble averaged measurements.

Measurement in general has no 100% efficiency due to decoherence, gate operation error and many more reasons. If this is the case, we have to repeat the same computation many times to achieve reasonably high reliability.

Moreover, we should be able to send and store quantum information to construct a quantum data processing network. This "networkability" requires following two additional criteria to be satisfied.

(6) *The ability to interconvert stationary and flying qubits.*

Some realizations are excellent in storing quantum information while long distant transmission of quantum information might require different physical resources. It may happen that some system has a Hamiltonian which is easily controllable and is advantageous in executing quantum algorithms. Compare this with a current digital computer, in which the CPU and the system memory are made of semiconductors while a hard disk drive is used as a mass storage device. Therefore a working quantum computer may involve several kinds of qubits and we are forced to introduce distributed quantum computing. Interconverting ability is also important in long distant quantum teleportation using quantum repeaters.

(7) *The ability to faithfully transmit flying qubits between specified locations.*

Needless to say, this is an indispensable requirement for quantum communication such as quantum key distribution. This condition is also important in distributed quantum computing mentioned above.

8.2. *Physical realizations*

There are numerous physical systems proposed as possible candidates for a viable quantum computer to date.[44] Here is the list of the candidates;

(1) Liquid-state/Solid-state NMR and ENDOR
(2) Trapped ions
(3) Neutral atoms in optical lattice
(4) Cavity QED with atoms
(5) Linear optics
(6) Quantum dots (spin-based, charge-based)
(7) Josephson junctions (charge, flux, phase qubits)
(8) Electrons on liquid helium surface

and other unique realizations. ARDA QIST roadmap[44] evaluates each of these realizations. The roadmap is extremely valuable for the identification and quantification of progress in this multidisciplinary field.

Acknowledgements

This summer school was supported by the "Open Research Center" Project for Private Universities: matching fund subsidy from MEXT (Ministry of Education, Culture, Sports, Science and Technology). Special thanks are due to the other organizers and coeditors of this lecture notes, Takashi Aoki, Robabeh Rahimi Darabad and Akira SaiToh.

References

1. M. Nakahara and T. Ohmi, *Quantum Computing: From Linear Algebra to Physical Realizations*, (Taylor and Francis, 2008).
2. M. A. Neilsen and I. L. Chuang, *Quantum Computation and Quantum Information*, (Cambridge University Press, 2000).
3. E. Rieffel and W. Polak, *ACM Computing Surveys (CSUR)* **32** (2000) 300.
4. Y. Uesaka, *Mathematical Principle of Quantum Computation*, (Corona Publishing, Tokyo, in Japanese, 2000).
5. P. A. M. Dirac, *Principles of Quantum Mechanics* (4th ed.), (Clarendon Press, 1981).
6. L. I. Shiff, *Quantum Mechanics* (3rd ed.), (McGraw-Hill, 1968).
7. A. Messiah, *Quantum Mechanics*, (Dover, 2000).
8. J. J. Sakurai, : *Modern Quantum Mechanics* (2nd Edition), (Addison Wesley, Boston, 1994).
9. L. E. Ballentine, *Quantum Mechanics*, (World Scientific, Singapore, 1998).
10. A. Peres, *Quantum Theory: Concepts and Methods*, (Springer, 2006).
11. A. SaiToh, R. Rahimi and M. Nakahara, e-print quant-ph/0703133.
12. A. Peres, *Phys. Rev. Lett.* **77** (1996) 1413.
13. M. Horodecki *et al.*, *Phys. Lett. A* **223** (1996) 1.
14. W. K. Wootters, and W. H. Zurek, *Nature* **299** (1982) 802.
15. M. A. Nielsen *et al.*, *Nature* **396** (1998) 52.
16. A. Barenco *et al.*, *Phys. Rev. A* **52** (1995) 3457.
17. Z. Meglicki, `http://beige.ucs.indiana.edu/M743/index.html`
18. D. Deutsch, *Proc. Roy. Soc. Lond. A*, **400** (1985) 97.
19. D. Deutsch and R. Jozsa, *Proc. Roy. Soc. Lond. A*, **439** (1992) 553.
20. E. Bernstein and U. Vazirani, *SIAM J. Comput.*, **26** (1997) 1411.
21. D. R. Simon, Proc. 35th Annual Sympo. Found. Comput. Science, (IEEE Comput. Soc. Press, Los Alamitos, 1994) 116.
22. T. Mihara and S. C. Sung, *Comput. Complex.* **12** (2003) 162.
23. M. A. Neilsen and I. L. Chuang, *Quantum Computation and Quantum Information*, (Cambridge University Press, 2000).
24. K. Hornberger, quant-ph/0612118.
25. H. Barnum, M. A. Nielsen and B. Schumacher, *Phys. Rev. A* **57** (1998) 4153.
26. Y. Kondo, *et al.*, *J. Phys. Soc. Jpn.* **76** (2007) 074002.
27. G. Lindblad, *Commun. Math. Phys.* **48** (1976) 119.
28. V. Gorini, A. Kossakowski and E. C. G. Sudarshan, *J. Math. Phys.*, **17** (1976) 821.
29. A. J. Fisher, Lecture note available at `http://www.cmmp.ucl.ac.uk/~ajf/course_notes.pdf`
30. A. M. Steane, quant-ph/0304016.
31. P. W. Shor, *Phys. Rev. A* **52** (1995) 2493.
32. A. Hosoya, *Lectures on Quantum Computation* (Science Sha, in Japanese, 1999).
33. F. J. MacWilliams and N. J. A. Sloane, *The Theory of Error-Correcting Codes*, (North-Holland, Amsterdam, 1977).
34. A. R. Calderbank and P. W. Shor, *Phys. Rev. A* **54** (1996) 1098.

35. A. M. Steane, *Phys. Rev. Lett.* **77** (1996) 793.
36. J Niwa, K. Matsumoto and H. Imai, quant-ph/0211071.
37. D. P. DiVincenzo and P. W. Shor, *Phys. Rev. Lett.* **77** (1996) 3260.
38. D. P. DiVincenzo, *Fortschr. Phys.* **48** (2000) 771.
39. M. Nakahara, S. Kanemitsu, M. M. Salomaa and S. Takagi (eds.) "Physical Realization of Quantum Computing: Are the DiVincenzo Criteria Fulfilled in 2004?" (World Scientific, Singapore, 2006).
40. J. Vartiainen *et al.*, *Phys. Rev. A* **70** (2004) 012319.
41. E. Knill, R. Laflamme, and L. Viola, Phys. Rev. Lett. **84**, 2525 (2000); P. Zanardi, *Phys. Rev. A* **63**, 012301 (2001); W. G. Ritter, *Phys. Rev. A* **72** (2005) 012305.
42. G. M. Palma, K. A. Suominen and A. K. Ekert, *Proc. R. Soc. London A* **452** (1996) 567; L. M. Duan and G. C. Guo, *Phys. Rev. Lett.* **79** (1997) 1953; P. Zanardi and M. Rasetti, *Phys. Rev. Lett.* **79** (1997) 3306; D. A. Lidar, I. L. Chuang, and K. B. Whaley, *Phys. Rev. Lett.* **81** (1998) 2594; P. Zanardi, *Phys. Rev. A* **60** (1999) 729(R); D. Bacon, D. A. Lidar, and K. B. Whaley, *Phys. Rev. A* **60** (1999) 1944.
43. D. P. DiVincenzo, *Phys. Rev. A* **51** (1995) 1015.
44. http://qist.lanl.gov/

BRAID GROUP AND TOPOLOGICAL QUANTUM COMPUTING

TAKAYOSHI OOTSUKA

Department of Physics, Kinki University, Osaka 577-8502, Japan
* E-mail: ootsuka@alice.math.kindai.ac.jp

KAZUHIRO SAKUMA

Department of Mathematics, Kinki University, Osaka 577-8502, Japan
* E-mail: sakuma@math.kindai.ac.jp

This is a survey article on the braid group and topological quantum computing. The purpose of this note is to discuss relation between a unitary representation of braid group and topological quantum computing.

Keywords: Symmetric Group, Braid Group, Presentation of a Group, Representation of a Group, Alexander Theorem, Markov Theorem, Jones Polynomial, Modular Functor, TQC.

1. Introduction

Historically in the process of development of theoretical physics, modern geometry has constantly provided a strong method and always played an important role at both fundamental and advanced level. For instance, it is obviously concluded from the facts that the notion of a Riemannian manifold, which is a direct generalization of surface theory, was employed as a model of the space universe, and that general relativity could be discussed as the Lorentz manifold theory. Moreover, it is impossible to discuss the gravitation field equation due to Einstein

$$R_{ij} - \frac{1}{2}Rg_{ij} + \Lambda g_{ij} = T_{ij}$$

without the language of Riemannian geometry. Also, we may say that based on the characterization and requirement from physics side, there has been much progress in the study of geometry closely related to manifold theory.

In the Erlangen program Felix Klein declared that "for one transformation group, there corresponds one geometry." This conception by Klein has been one of the most important teaching principles in modern geometry. More precisely, for a given space X and a transformation group G, we can study invariant properties of a subset A in X under a transformation in G, which is called a *geometry of X belonging to G*. For example, the Euclidean geometry is a branch to study invariant properties of a figure A in $X = \mathbb{R}^2$ under congruent transformation or similar transformation. In its simplest form it is the Side-Angle-Side theorem of middle school geometry which characterizes triangles up to congruence or the Angle-Angle theorem up to similarity. In addition, differential topology is a branch to study invariant properties of a manifold up to diffeomorphism or homeomorphism. Thus from a modern mathematical viewpoint, geometry is a mathematics in which one searches for invariant properties of not only figures but also functions on the figures or maps between figures under an element of a group G.

The ultimate purpose of group theory which is one of the branches in algebra is to give a group structure for any set and classify all groups up to isomorphism. Group theory was established by Galois with an excellent idea in order to solve an equation systematically. It gave not merely a solution of an equation but an answer to a geometrical problem (e.g. a regular n-polygon is of possible construction if and only if the Galois group [a] $\mathrm{Gal}(\mathbb{Q}(e^{\frac{\pi i}{n}})/\mathbb{Q} \cong (\mathbb{Z}/2n\mathbb{Z})^\times$ has order 2^r). Thus when we consider a problem basically related with group theory, we see that it is tied up with geometry. Conversely, a group naturally appears in considering a geometrical problem. This is because geometry is a mathematics in which one solves equations on a geometrical object.

On one hand, a group often appears also in physics in a natural way. As is seen in § 4, braid groups are used in describing topological quantum computation, which is a recent work by M. H. Freedman, M. Larsen and Z. Wang. So, in the next section we discuss braid groups together with basic facts on group theory. For readers' convenience, we provide an appendix where we discuss basic facts on the fundamental groups including its definition.

Acknowledgement. The authors wishes to thank Mikio Nakahara for well organizing summer school held at August 2007 and stimulating discussion. He also wishes to thank the participants at the summer school for

---

[a] Here we do not give a definition of Galois group because we do not need it in the later discussion

their useful comments and inspiring questions. The second author would like to thank his student Kumi Kobata for her help in writing this paper.

2. Braid Groups

In this section we review basic concepts of braid group mainly from an algebraic viewpoint, which will be a quick course for physicists. First we start with recalling the definition of a group and fundamental concepts in group theory:

Definition 2.1. For any set G, if a map (binary operation) $\mu : G \times G \to G$ satisfies the following three conditions, then a pair (G, μ) or simply G is called a *group*.

(1) For any a, b, $c \in G$ we have

$$\mu(a, \mu(b, c)) = \mu(\mu(a, b), c).$$

(2) For any $a \in G$ there exists an element $e \in G$ such that

$$\mu(a, e) = \mu(e, a) = a,$$

where e is called a *unit element* of G.

(3) For any $a \in G$ there exists an element $a' \in G$ such that

$$\mu(a, a') = \mu(a', a) = e,$$

where a' is called an *inverse element* of a and we write $a^{-1} \in G$.

In the above definition a binary operation μ is called a *product* and it is usually written as $a \cdot b \in G$ or $ab \in G$ in place of $\mu(a, b)$ for simplicity. Note that we can make a set G to be a group with different products in general.

Given a group (G, μ) if we have $\mu(a, b) = \mu(b, a)$ for any a, $b \in G$, then we call G an *abelian* group. Let H be a subset of a group (G, μ). If H is a group under the product μ of G, H is called a *subgroup* of G. Note that a subset H is a subgroup of G if and only if $\mu(a, b^{-1}) \in H$ for any a, $b \in H$.

Let G be a group and H a subgroup of G. Then, for any $g \in G$ we define two subgroups as follows:

$$gH = \{gh;\ h \in H\}, \qquad Hg = \{hg;\ h \in H\}.$$

If $gH = Hg$ for any $g \in G$, then H is called a *normal subgroup* of G. Note that H is a normal subgroup of G if and only if $g^{-1}hg \in H$ for any $g \in G$ and any $h \in H$.

Let H be a normal subgroup of G. Then we define $g_1 \sim g_2$ if and only if $g_1 g_2^{-1} \in H$. This relation $\sim$ is an equivalence relation; $gg^{-1} = e \in H$ implies $g \sim g$, $(g_1 g_2^{-1})^{-1} = g_2 g_1^{-1} \in H$ implies $g_2 \sim g_1$ if we suppose that $g_1 \sim g_2$, and $(g_1 g_2^{-1})(g_2 g_3^{-1}) = g_1 g_3^{-1}$ implies $g_1 \sim g_3$ if we suppose that $g_1 \sim g_2$, $g_2 \sim g_3$. We denote the quotient group under this relation by G/H. For any $g \in G$ the equivalence class of g is denoted by $[g] = gH = Hg \in G/H$, where g is called a representative. For $[g_1]$, $[g_2] \in G/H$ the product is defined as follows:

$$[g_1][g_2] = [g_1 g_2] \in G/H.$$

Here note that the product is well-defined, i.e., the product $[g_1 g_2]$ does not depend on the choices of representatives g_1, g_2.

Example 2.1.

(1) Let $\mathbb{R}$ denote the set of real numbers. It is easy to see that $\mathbb{R}$ is a group with a product defined by $\mu(a, b) = a + b$. Then we say that $(\mathbb{R}, \mu)$ is an *additive group*. However, $(\mathbb{R}, \mu')$ is *not* a group when we employ another operation $\mu'(a, b) = ab$. This is beacause there exists a unit $1 \in \mathbb{R}$ but there is no inverse for $0 \in \mathbb{R}$. The n-dimensional linear space $\mathbb{R}^n$ is naturally a group with $\mu(\boldsymbol{x}, \boldsymbol{y}) = \boldsymbol{x} + \boldsymbol{y}$ for any $\boldsymbol{x}$, $\boldsymbol{y} \in \mathbb{R}^n$.

(2) Let $\mathbb{Z}$ denote the set of integers. Then, $\mathbb{Z}$ is a group with $\mu(a, b) = a + b$ and a subgroup of $(\mathbb{R}, \mu)$. But $(\mathbb{Z}, \mu')$ is not a group with $\mu'(a, b) = ab$. The direct product $\mathbb{Z} \oplus \mathbb{Z} = \{(x, y);\ x, y \in \mathbb{Z}\}$ is a group with $\mu((x_1, y_1), (x_2, y_2)) = (x_1 + x_2, y_1 + y_2)$.

(3) Let $K = \mathbb{Q}$ (the set of rational numbers), $\mathbb{R}$, or $\mathbb{C}$ (the set of complex numbers). Then, $(K, +)$ is an additive group similarly as in (1). Furthermore, it is easy to see that $K - \{0\}$ is a group with a product defined by $\mu(a, b) = ab$, which is called a *multiplicative* group.

(4) Let n be a natural number. It is easy to see that $n\mathbb{Z} = \{nm;\ m \in \mathbb{Z}\}$ is a normal subgroup of $\mathbb{Z}$. Consider the quotient group $\mathbb{Z}/n\mathbb{Z}$. Then, $[a] = \{a + nm;\ m \in \mathbb{Z}\}$ and $[a] + [b] = [a + b] = \{a + b + nm;\ m \in \mathbb{Z}\}$ is well-defined. It is often written as $\mathbb{Z}/n\mathbb{Z} = \mathbb{Z}_n = \{[0], [1], \ldots, [n-1]\}$.

In order to compare the structures of given groups, we need to define a map which preserves the products.

Definition 2.2. Let (G, μ) and (G', μ') be groups. If a map $\varphi : G \to G'$ preserves the products, i.e.,

$$\varphi(\mu(a, b)) = \mu'(\varphi(a), \varphi(b)),$$

then φ is called a *homomorphism*[b]. In other words, we have the following commutative diagram:

$$
\begin{array}{ccc}
G \times G & \xrightarrow{\ \varphi \times \varphi\ } & G' \times G' \\
\mu \downarrow & \circlearrowleft & \downarrow \mu' \\
G & \xrightarrow[\ \varphi\ ]{} & G' .
\end{array}
$$

Let e' be a unit element of G'. Given a homomorphism $\varphi : G \to G'$, we can define a set

$$\mathrm{Ker}(\varphi) = \{g \in G;\ \varphi(g) = e'\},$$

which is called the *kernel* of φ. Moreover, we can define a set

$$\mathrm{Im}(\varphi) = \varphi(G) = \{\varphi(g) \in G';\ \forall g \in G\},$$

which is called the *image* of φ. Note that $\mathrm{Ker}(\varphi)$ is a normal subgroup of G and $\mathrm{Im}(\varphi)$ a subgroup of G'.

If a homomorphism $\varphi : G \to G'$ is bijective, i.e.,
(1) if $\varphi(a) = \varphi(b)$, then $a = b$, and
(2) $\mathrm{Im}(\varphi) = G'$,
then φ is called an *isomorphism*. We say that G is *isomorphic* to G' if there exists an isomorphism $\varphi : G \to G'$, denoted by $G \cong G'$. When (1) holds, φ is said to be *injective*. When (2) holds, φ is said to be *surjective*. In group theory, two different sets are identified even if they have different products as groups but there exists an isomorphism between them.

Given a homomorphism $\mu : G \to G'$, we can consider a quotient group $G/\mathrm{Ker}(\varphi)$. Any element of $G/\mathrm{Ker}(\varphi)$ is denoted by $[g]$ as an equivalence class of $g \in G$. Then we define a map $f : G/\mathrm{Ker}(\varphi) \to \mathrm{Im}(\varphi)$ by $f([g]) = \varphi(g)$. It is easy to see that the map f is a well-defined homomorphism and bijective, which means that $G/\mathrm{Ker}(\varphi) \cong \mathrm{Im}(\varphi)$.

Exercise 2.1. Let $\varphi : G_1 \to G_2$ be a homomorphism.

---

[b]If we omit the notation of the products, the definition of a homomorphism is simply given as follows: $\varphi(ab) = \varphi(a)\varphi(b)$.

(1) Show that if $e \in G_1$ is an unit element, then e uniquely exists and $\varphi(e) \in G_2$ is also an unit element.

(2) Show that $\varphi(g^{-1}) = \{\varphi(g)\}^{-1}$.

(3) Prove that $G_1/\mathrm{Ker}(\varphi) \cong \mathrm{Im}(\varphi)$.

(4) Show that if φ is an isomorphism, then φ^{-1} is also an isomorphism.

(5) Suppose that there exists a homomorphism $\psi : G_2 \to G_1$ such that $\psi \circ \varphi = \mathrm{id}_{G_1}$ and $\varphi \circ \psi = \mathrm{id}_{G_2}$. Then, show that φ is an isomorphism.

Definition 2.3. Let G be a group. If there are a finite (or infinite) number of elements g_1, g_2, ..., g_k, ... $\in G$ such that an arbitrary element of $g \in G$ can be expressed by a finite (or infinite) number of products of those g_1, g_2, ..., g_1^{-1}, g_2^{-1}, ..., then we say that g_1, g_2, ... are *generators* of G and that G is *generated by* those generators. If G is generated by a finite number of generators, then G is said to be *finitely generated*.

Now we consider a sequence with a finite number of elements; $a, b, \ldots, n$. Such a sequence consisting of $a, b, \ldots, n$ is called a *word*. For example,

$$aa, \ abc, \ aabbcc, \ ababab, \ ccc, \ cab$$

are words defined by three elements a, b, c. For two words x and y, we define the product xy as its juxtaposition, e.g. for $x = abc$, $y = cab$ we have $xy = abccab$. We can consider an empty word " "or e, and the inverses a^{-1}, b^{-1}, c^{-1} for a, b, c, i.e., $aa^{-1} = e$ (empty word). We employ, namely, such a (economical) rule that if a given word includes an empty word, then we ignore it like the following:

$$abbb^{-1}c = abec = ab \ c = abc.$$

Thus we should regard that $abbb^{-1}c$ and abc are same words under this rule. We call this an *irreducible word*. It is easy to see that the set of all irreducible words consisting of n elements make a group under the above product. Note that an empty word is a unit.

Definition 2.4. The set of all irreducible words consisting of distinct $2n$ elements

$$a_1, \ a_2, \ldots, \ a_n, \ a_1^{-1}, \ a_2^{-1}, \ldots, \ a_n^{-1}$$

with a group structure under the above product is called a *free group with rank* n, denoted by F_n or $F(a_1, \ldots, a_n)$. Then, $a_1, \ldots, a_n$ are generators of F_n. For simplicity, for any $x \in F_n$ we employ the notation on k times product of x in the following; $x^k = \underbrace{x \cdots x}_{k}$.

Let G and H be groups. We call a word again an element which consists of a finite number of arbitrarily juxtaposed elements of G and H. For example,

$$g_1 h_1 g_2 g_3 h_2 h_3 h_4 \qquad (g_i \in G, \; h_i \in H)$$

is a word consisting of 7 elements. Then we introduce a rule such that if there are adjacent elements which belong to a same group, then we replace them by their products. For example, in the above word,

$$g_1 h_1 g_2 g_3 h_2 h_3 h_4 \qquad \Longrightarrow \qquad g_1 h_1 (g_2 \circ g_3)(h_2 \circ h_3 \circ h_4),$$

where we regard the latter as the word consisting of 4 elements. Under this rule, it is easy to see that the set of all words consisting of G and H forms a group. We denote this group by $G * H$ or $H * G$, which is called the *free product* of G and H.

Let G be a finitely generated group with n generators $g_1, \ldots, g_n$. Then we consider a free group $F_n = F(a_1, \ldots, a_n)$ with rank n and define a map $f : F_n \to G$, $f(a_i) = g_i$ such that f is a homomorphism, eg. $f(a_1 a_2^{-1} a_3) = g_1 g_2^{-1} g_3$. Clearly, f is a surjection, i.e., $f(F_n) = G$. Then, by Exercise 2.1 (3), we have

$$F_n/\mathrm{Ker}(f) \cong \mathrm{Im}(f) = G,$$

which means that any finitely generated group G is isomorphic to a quotient group of a free group.

Suppose that the subgroup $\mathrm{Ker}(f)$ is generated by $h_1, \ldots, h_r$, denoted by $\mathrm{Ker}(f) = (h_1, \ldots, h_r)$. Thus we have the isomorphic correspondence

$$F(a_1, \ldots, a_n)/(h_1, \ldots, h_r) \cong G, \qquad (1)$$

which is called a *presentation* of G and $a_1, \ldots, a_n$ are called generators, and $h_1, \ldots, h_r$ *relations*. We usually express this like the following:

$$G = \langle a_1, \ldots, a_n | \; h_1 = e, \ldots, \; h_r = e \rangle.$$

Note that the way of the presentation of a given G is not unique in general. As for a free group, we have

$$F_n = \langle a_1, \ldots, a_n | \qquad \rangle$$

since it has no relations, but we usually write it in the following: $F_n \cong \langle a_1, \ldots, a_n \rangle$.

Example 2.2. Let us find the presentation of $\mathbb{Z}_3 = \{[0], [1], [2]\}$, which is a finite additive group of order 3. Let $F(a)$ denote a free group with rank 1 which is canonically isomorphic to $\mathbb{Z}$. Then we define a surjective homomorphism $f : F(a) \to \mathbb{Z}_3$, $f(a) = [1]$. Thus we can verify that

$$f(a^2) = f(a) + f(a) = [1] + [1] = [2]$$
$$f(a^3) = f(a) + f(a) + f(a) = [1] + [1] + [1] = [0]$$
$$f(a^4) = [4] = [1], \text{etc.}$$

Hence we have $\mathrm{Ker}(f) = \{a^{3m};\ m \in \mathbb{Z}\}$, which is generated by a^3. Therefore we have a presentation $\mathbb{Z}_3 \cong F(a)/(a^3) \cong \langle a|\ a^3 = e \rangle$.

In general, for any natural number n we have a presentation

$$\mathbb{Z}_n \cong \langle a|\ a^n = e \rangle.$$

Example 2.3. $F(a, b) \cong \mathbb{Z} * \mathbb{Z}$. In general, for $G = F(a_1, \ldots, a_m)$ and $H = F(b_1, \ldots, b_n)$ we have

$$G * H = F(a_1, \ldots, a_m, b_1, \ldots, b_n).$$

Exercise 2.2. Show that the following representation is true;

$$\mathbb{Z} \oplus \mathbb{Z} \cong \langle a, b|\ aba^{-1}b^{-1} = e \rangle.$$

Here let us recall the definition of a symmetric group which is closely related to a braid group. Let $X_n = \{1, 2, \ldots, n\}$ be a finite set consisting of n elements. Then we define a set

$$\mathfrak{S}_n = \{\sigma : X_n \to X_n, \text{bijection}\},$$

which is a group with a product defined by composing maps. $\mathfrak{S}_n$ is called an *n-th symmetric group*, whose order is $n!$. Any element $\sigma \in \mathfrak{S}_n$ is often written as follows:

$$\sigma = \begin{pmatrix} 1 & 2 & \cdots & n \\ \sigma(1) & \sigma(2) & \cdots & \sigma(n) \end{pmatrix},$$

and then, for $\tau, \sigma \in \mathfrak{S}_n$, the product is

$$\tau \circ \sigma = \begin{pmatrix} 1 & 2 & \cdots & n \\ \tau(\sigma(1)) & \tau(\sigma(2)) & \cdots & \tau(\sigma(n)) \end{pmatrix}.$$

Also, $e = \begin{pmatrix} 1\ 2\ \cdots\ n \\ 1\ 2\ \cdots\ n \end{pmatrix}$ is a unit element of $\mathfrak{S}_n$.

For example, we immediately see that $\mathfrak{S}_2 = \{e, \sigma\} \cong \langle a|\, a^2 = e \rangle \cong \mathbb{Z}_2$ if we define $f : F(a) \to \mathfrak{S}_2$ by $f(a) = \sigma$, where $\sigma = \begin{pmatrix} 1 & 2 \\ 2 & 1 \end{pmatrix}$ and note that $\sigma \circ \sigma = e$. Moreover, we have $\mathfrak{S}_3 = \{e, \sigma_1, \sigma_2, \sigma_3, \tau_1, \tau_2\}$, where

$$\sigma_1 = \begin{pmatrix} 1 & 2 & 3 \\ 1 & 3 & 2 \end{pmatrix}, \sigma_2 = \begin{pmatrix} 1 & 2 & 3 \\ 3 & 2 & 1 \end{pmatrix}, \sigma_3 = \begin{pmatrix} 1 & 2 & 3 \\ 2 & 1 & 3 \end{pmatrix}, \tau_1 = \begin{pmatrix} 1 & 2 & 3 \\ 2 & 3 & 1 \end{pmatrix}, \tau_2 = \begin{pmatrix} 1 & 2 & 3 \\ 3 & 1 & 2 \end{pmatrix}.$$

It is easy to check that

$$\sigma_1 \circ \sigma_2 = \tau_1, \quad \sigma_2 \circ \sigma_1 = \tau_2, \quad \sigma_1 \circ \sigma_2 \circ \sigma_1 = \sigma_2 \circ \sigma_1 \circ \sigma_2 = \sigma_3. \tag{2}$$

Note also that $\sigma_1 \circ \sigma_1 = \sigma_2 \circ \sigma_2 = e$. Thus we have seen that $\mathfrak{S}_3$ is generated by σ_1 and σ_2, and it is not apparently abelian.

Let us find a presentation of $\mathfrak{S}_3$. We define a surjective homomorphism $f : F(a, b) \to \mathfrak{S}_3$ as follows:

$$f(a) = \sigma_1, \quad f(b) = \sigma_2.$$

Then, we see that $f(a^2) = \sigma_1 \circ \sigma_1 = e$, $f(b^2) = \sigma_2 \circ \sigma_2 = e$ and $f((ab)^3) = (\sigma_1 \circ \sigma_2)^3 = (\tau_1)^3 = e$. Thus f induces a surjective homomorphism

$$\tilde{f} : F(a, b)/(a^2, b^2, (ab)^3) \to \mathfrak{S}_3$$

Hence in order to prove that $\tilde{f}$ is an isomorphism, it suffices to show that $F(a, b)/(a^2, b^2, (ab)^3)$ has order 6, which means that $\tilde{f}$ is injective since $\mathfrak{S}_3$ also has order 6. We will leave the details to the reader as an easy exercise.

Thus, we have a presentation

$$\mathfrak{S}_3 \cong \langle a, b \mid a^2 = e,\ b^2 = e,\ (ab)^3 = e \rangle.$$

Here we shall put $s_1 = a$, $s_2 = bab$. Then, $s_1^2 = a^2 = e$, $s_2 = babbab = e$, and $s_1 s_2 s_1 = ababa = (ab)^3 b^{-1} = b$, $s_2 s_1 s_2 = bababab = b(ab)^3 = b$. Thus we have another presentation

$$\mathfrak{S}_3 \cong \langle s_1, s_2 \mid s_1^2 = e,\ s_2^2 = e,\ s_1 s_2 s_1 = s_2 s_1 s_2 \rangle.$$

In general we have

$$\mathfrak{S}_n = \left\langle s_1, \ldots, s_{n-1} \,\middle|\, \begin{array}{ll} s_i^2 = 1 & (i = 1, \ldots, n-1) \\ s_i s_{i+1} s_i = s_{i+1} s_i s_{i+1} & (i = 1, \ldots, n-2) \\ s_i s_j = s_j s_i & (|i - j| \geq 2) \end{array} \right\rangle,$$

which is not abelian for $n \geq 3$ since $s_i s_{i+1} \neq s_{i+1} s_i$.

Exercise 2.3. Consider a set $G = \{E,\ A,\ B,\ C,\ D,\ F\}$, where

$$E = \begin{pmatrix} 1 & 0 & 0 \\ 0 & 1 & 0 \\ 0 & 0 & 1 \end{pmatrix},\ A = \begin{pmatrix} 1 & 0 & 0 \\ 0 & 0 & 1 \\ 0 & 1 & 0 \end{pmatrix},\ B = \begin{pmatrix} 0 & 1 & 0 \\ 1 & 0 & 0 \\ 0 & 0 & 1 \end{pmatrix},$$

$$C = \begin{pmatrix} 0 & 0 & 1 \\ 0 & 1 & 0 \\ 1 & 0 & 0 \end{pmatrix}, \ D = \begin{pmatrix} 0 & 0 & 1 \\ 1 & 0 & 0 \\ 0 & 1 & 0 \end{pmatrix}, \ F = \begin{pmatrix} 0 & 1 & 0 \\ 0 & 0 & 1 \\ 1 & 0 & 0 \end{pmatrix}.$$

(1) Show that G is a group by matrix multiplication.
(2) Show that $A^2 = B^2 = C^2 = E$, $F^2 = D$, $AD = C$, $F^3 = E$ and hence $G = \{A, A^2, F, F^2, AF, AF^2\}$.
(3) Show that G is isomorphic to $\mathfrak{S}_3$.

3. Knots Defined by Braids

Let K be a disjoint union of circles, i.e., $K = S^1 \cup \cdots \cup S^1$, where S^1 is a circle. An embedding of K into the 3-dimensional Euclidean space $\mathbb{R}^3$ (or the 3-sphere S^3), $\varphi : K \to \mathbb{R}^3$, is called a *link*; in particular, φ is a *knot* if K is connected, i.e., $K = S^1$. Given two embeddings $\varphi_1, \varphi_2 : K \to \mathbb{R}^3$, we say that φ_1 is *ambiently isotopic* to φ_2 if there exist homeomorphisms $h : K \to K$ and $H : \mathbb{R}^3 \to \mathbb{R}^3$ such that $H \circ \varphi_1 = \varphi_2 \circ h$, i.e., there is the following commutative diagram:

$$\begin{array}{ccc} K & \xrightarrow{\ \ h\ \ } & K \\ {\scriptstyle \varphi_1} \downarrow & \circlearrowleft & \downarrow {\scriptstyle \varphi_2} \\ \mathbb{R}^3 & \xrightarrow[\ \ H\ \ ]{} & \mathbb{R}^3 \end{array}$$

Let $\mathcal{L}$ be the set of all links. Suppose that a map $\lambda : \mathcal{L} \to \Lambda$ is defined for an appropriately calculable algebra Λ. Such a map λ that if K_1 is ambiently isotopic to K_2, then $\lambda(K_1) = \lambda(K_2)$ for any K_1, $K_2 \in \mathcal{L}$ is called an *invariant* of links. Many invariants have been discovered and we will introduce one of them, the Jones polynomial, later on. The ultimate purpose in knot theory is to find the complete invariant in which the converse implication holds. Unfortunately, it has not been discovered yet.

Now we recall the definition of braid group which plays an important role in knot theory. First we take $2n$ points in a cube $C = [0, 1] \times [0, 1] \times [0, 1]$ which is a subset of $\mathbb{R}^3$ in the following:

$$P_1 = \left(\frac{1}{2}, \frac{1}{n+1}, 1 \right), \ldots, P_n = \left(\frac{1}{2}, \frac{n}{n+1}, 1 \right)$$

$$P_1' = \left(\frac{1}{2}, \frac{1}{n+1}, 0 \right), \ldots, P_n' = \left(\frac{1}{2}, \frac{n}{n+1}, 0 \right).$$

Then we join the P_1, $P_2, \ldots,$ P_n to P'_1, $P'_2, \ldots,$ P'_n by n arcs in C in such a way that these n arcs do not mutually intersect each other and each arc is not self-knotted. We say that these n curves in C are an n-braid. If we parametrize the arc γ by $(x(t), y(t), t)$, then γ intersects with $z = t$ plane at one and only one point for any $t \in [0, 1]$. Given two n-braids in C, if we can transform one to the other by performing the elementary knot moves on these arcs, then we say that these two n-braids are equal.

Let B_n be the set of all n-braids. For $\sigma, \tau \in B_n$ we define a product of σ and τ as follows: First glue the base of the cube containing σ to the top face of the cube containing τ. As a resultant, we have a rectangular solid in which there are n-arcs composed by the vertical juxtaposition of σ and τ. By shrinking the rectangular solid in half we can obtain an n-braid which is the product of σ and τ, denoted by $\sigma\tau$. Note that in general $\sigma\tau$ is not equal to $\tau\sigma$, i.e., $\sigma\tau \neq \tau\sigma$. We can make B_n a group under this product and call it the n-th barid group. Let $e \in B_n$ be an n-braid obtained by joining the P_1, $P_2, \ldots,$ P_n to P'_1, $P'_2, \ldots,$ P'_n respectively by n segments (see Fig. 1), which is called a *trivial* n-braid and a unit element in B_n.

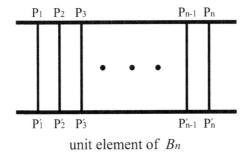

unit element of B_n

Fig. 1.

The braid group B_n can be represented by generators and defining relations between the generators. In order to give a presentation we shall introduce two specific n-braids, say σ_1, σ_i^{-1}, which connect P_i to P'_{i+1} and P_{i+1} to P'_i, and then connect the remaining P_j to P'_j for $j \neq i, i+1$ by segments:

It is shown that any n-braid can be represented as the product of the elements $\sigma_i^{\pm}$ for $i = 1, 2, \ldots, n-1$. Thus the elements $\sigma_1, \ldots,$ σ_{n-1} are generators of B_n.

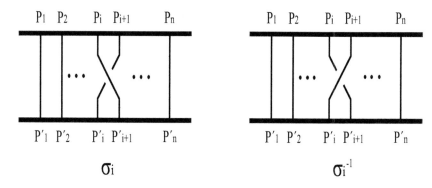

Fig. 2.

For example, B_2 is generated by σ_1 or σ_1^{-1}. Any element $\alpha \in B_2$ is uniquely determined by the number of σ_1's minus the number of σ_1^{-1}'s in α. Thus $B_2 = \langle \sigma_1 \rangle$, which is isomorphic to a free group $\mathbb{Z}$. Moreover, we have

$$B_3 = \langle \sigma_1, \sigma_2 \mid \sigma_1 \sigma_2 \sigma_1 = \sigma_2 \sigma_1 \sigma_2 \rangle$$
$$B_4 = \langle \sigma_1, \sigma_2, \sigma_3 \mid \sigma_1 \sigma_3 = \sigma_3 \sigma_1, \ \sigma_1 \sigma_2 \sigma_1 = \sigma_2 \sigma_1 \sigma_2, \ \sigma_3 \sigma_2 \sigma_3 = \sigma_2 \sigma_3 \sigma_2 \rangle .$$

In general, we have the following presentation:

$$B_n = \left\langle \sigma_1, \ldots, \sigma_{n-1} \ \middle| \ \begin{matrix} \sigma_i \sigma_{i+1} \sigma_i = \sigma_{i+1} \sigma_i \sigma_{i+1} & (i = 1, \ldots, n-2) \\ \sigma_i \sigma_j = \sigma_j \sigma_i & (|i-j| \geq 2) \end{matrix} \right\rangle ,$$

which is due to Artin (see eg. p. 51 in Prasolov-Sossinsky[8]). Thus we have a (natural) surjective homomorphism $\Psi : B_n \to \mathfrak{S}_n$ defined by $\Psi(\sigma_i) = s_i$ for $i = 1, 2, \ldots, n-1$, which just corresponds to forgetting the (under- or over-) crossing in σ_i for any i.

The braid group naturally appears in geometrical or topological context. For example, if we set $S = \{(x, y, z) \in \mathbb{R}^3; x^2 + y^2 = 1, \ z = 0\}$ which is a trivial knot in $\mathbb{R}^3$, then we have $\pi_1(\mathbb{R}^3 - S) \cong \mathbb{Z} \ (\cong B_2)$. If K represents a trefoil knot in $\mathbb{R}^3$ (see Fig. 3), then we have $\pi_1(\mathbb{R}^3 - K) \cong B_3$. (See Example 1.24 in Hatcher,[5] note that a trefoil knot is a (2,3)-torus knot.)

Let σ be any n-braid in B_n. By means of a set of parallel n arcs which lie outside the cube C, we can connect the n points $P_1, \ldots, P_n$ on the top face of σ to the n points $P'_1, \ldots, P'_n$ respectively on the bottom face. Then we have obtained a regular diagram of a knot or link from the n-braid σ. A knot or a link constructed in this way is said to be a *closed braid* (or a

trefoil K

Fig. 3.

closure of σ). For example, we have a trefoil knot as a closure of 3-braid σ_1^3 and a figure eight knot as that of 4-braid $\sigma_1 \sigma_2^{-1} \sigma_1 \sigma_2^{-1}$.

The usefulness of this definition is based on the following fact due to Alexander:

Theorem 3.1 (the Alexander theorem). *Any knot or link can be obtained as a closed braid.*

As for the proof, see eg. p. 55 in Prasolov-Sossinsky.[8] Note that we may have such a situation that a knot or link is obtained by distinct braids. For example, we have a trivial knot as the closure of 2-braid σ_1, the 3-braid $\sigma_1^{-1} \sigma_2$ and the 4-braid $\sigma_1 \sigma_2 \sigma_3^{-1}$.

We define two operations in the union of braid groups which are called *Markov moves*:

(1) **(Markov move I)** For $b \in B_n$, transform b into xbx^{-1} for some $x \in B_n$.
(2) **(Markov move II)** For $b \in B_n$, transform b into either of two $(n+1)$-braids $b\sigma_n$ or $b\sigma^{-1}$ in B_{n+1}.

Definition 3.1. Let a, $b \in \mathcal{B} = \bigcup_{n \geq 1} B_n$. If we can transform a into b by a finite number of performing the Markov moves I, II defined above and their inverses, then a is said to be *M-equivalent* to b, denoted by $a \overset{M}{\sim} b$.

Theorem 3.2 (the Markov theorem). *Let K_1 and K_2 be knots (or links) obtained as the closure of a_1 and a_2 respectively in $\mathcal{B}$. Then, K_1 is ambiently isotopic to K_2 if and only if $a_1 \overset{M}{\sim} a_2$.*

See the proof in Birman.[2] Let $\mathcal{L}$ be the set of ambient isotopy classes of oriented links. Theorem 3.2 claims that a map $\Psi : \mathcal{B}/\overset{M}{\sim} \to \mathcal{L}$ is a bijection, which enables us to classify all knots or links by finding algebraic operations which should be invariant under the Markov moves I and II.

Next we shall discuss the representation of a group. Let V be a vector space over a field. Let $GL(V)$ be the set of all linear maps of V into V which are regular. $GL(V)$ is a group with a product as a composite of linear maps, i.e., $f \cdot g = f \circ g \in GL(V)$ for f, $g \in GL(V)$. For a group G a homomorphism $\varphi : G \to GL(V)$ is called a *(linear) representation* of G, V a *representation space* and the dimension of V the *dimension* of the representation.

Example 3.1. Let us find a one dimensional representation of B_n. Note that $GL(\mathbb{C}) \cong \mathbb{C} - \{0\}$ and then the relation $\sigma_i \sigma_j = \sigma_j \sigma_i$ is trivially satisfied because $GL(\mathbb{C})$ is an abelian group. Also, from the relation $\sigma_i \sigma_{i+1} \sigma_i = \sigma_{i+1} \sigma_i \sigma_{i+1}$ we can deduce that $\sigma_1 = \cdots = \sigma_{n-1}$. Hence, for any $z \in GL(\mathbb{C})$, the correspondence $f(\sigma_i) = z$ $(i = 1, \ldots, n-1)$ gives a representation of B_n. For $\forall b \in B_n$ we define $\varepsilon(b)$ to be the number of σ_i's in b minus that of σ_i^{-1}'s in b for any i, which is called a *crossing number* of $b \in B_n$. Thus we have obtained a one dimensional representation $f_z : B_n \to GL(\mathbb{C})$ defined by $f_z(b) = z^{\varepsilon(b)}$. If we restrict to $z \in U(1) = \{z \in \mathbb{C}; |z| = 1\}$, we have a one dimensional unitary representation $f_z : B_n \to U(1)$.

Let $V = \mathbb{C}^2$ and then we consider the tensor space $V \otimes V$. Take an element

$$
T = \begin{pmatrix} 1 & 0 & 0 & 0 \\ 0 & 0 & -q & 0 \\ 0 & -q & 1-q^2 & 0 \\ 0 & 0 & 0 & 1 \end{pmatrix} \in GL(V \otimes V),
$$

where $q \in \mathbb{C} - \{0\}$. Moreover, we consider its n times tensor space $V^{\otimes n} = \underbrace{V \otimes \cdots \otimes V}_{n}$ and set

$$
T_i = \underbrace{I \otimes \cdots \otimes I}_{i-1} \otimes T \otimes \underbrace{I \otimes \cdots \otimes I}_{n-i-1} \in GL(V^{\otimes n}),
$$

where I stands for the identity map of V. Then it is easy to verify that $T_1, T_2, \ldots, T_{n-1}$ satisfy the relations in B_n;

$$
T_i T_j = T_j T_i, \quad T_i T_{i+1} T_i = T_{i+1} T_i T_{i+1}.
$$

So, we define $\rho_n : B_n \to GL(V^{\otimes n})$ by $\rho_n(\sigma_i) = T_i$ for each generator $\sigma_i \in B_n$ and then we easily see that ρ_n is a representation of B_n.

Suppose that a representation of the n-th braid group $f_n : B_n \to GL(V_n)$ is given for any n. This induces a map $f : \bigcup_{n \geq 1} B_n \to \bigcup_{n \geq 1} GL(V_n)$. For any $b \in \mathcal{B} = \bigcup_{n \geq 1} B_n$ consider its trace $\mathrm{Tr}(f(b))$ of a matrix $f(b)$. Since f is a homomorphism, we have

$$\mathrm{Tr}(f(xbx^{-1})) = \mathrm{Tr}(f(x)f(b)f(x^{-1})) = \mathrm{Tr}(f(x)f(b)f(x)^{-1}) = \mathrm{Tr}(f(b)),$$

which means that the correspondence $f(b) \mapsto \mathrm{Tr}(f(b))$ is invariant under the Markov move I. With a little more correction, we obtain a correspondence which is invariant both under the Markov moves I and II. For any link K we take an n-braid b. Then we define

$$P_K(q) = q^{\varepsilon(b)}\mathrm{Tr}(\rho_n(b)h^{\otimes n}),$$

where $\varepsilon(b)$ is the crossing number already defined in Example 3.1, $h = \begin{pmatrix} q^{-1} & 0 \\ 0 & q \end{pmatrix}$ and $h^{\otimes n} = \underbrace{h \otimes \cdots \otimes h}_{n}$. We can check that $P_K(q)$ is a Laurant polynomial with a variable q and invariant under the Markov moves I and II. $P_K(q)$ is called the *Jones polynomial* of K (see Jones[6]).

4. Topological Quantum Computing

Roughly speaking, quantum computing is a unitary transformation in the process of computing (Nakahara[7]). Recalling that a representation of a group is a homomorphism into general linear space, i.e., a linear transformation will be assigned for an element of a group, we may regard that if the representation will be realized as a unitary transformation, then knots or links express its process of computation in quantum computing by replacing a unitary transformation in quantum computing with a representation of the braid group B_n.

Recently, Freedman-Kitaev-Larsen-Wang[4] introduced the notion of a 'modular functor', which is an extension of a representation of the braid group. They have shown that a unitary modular functor is equivalent to quantum computation. More precisely, given a problem in BQP, a linear map which corresponds to a self-homeomorphism of a disk with holes can be approximated by a sequence of quantum gates for an arbitrary unitary modular functor. Conversely, any quantum gate can be approximated by a sequence of self-homeomorphisms. Here, BQP is the class of decision problems which can be solved with probability being greater than or equal to 0.75 by an exact quantum circuit designed by a classical algorithm in $poly(L)$, where L is the length of the problem instance M.

We shall give a definition of a modular functor. Let X be a compact oriented surface with boundaries. Suppose that each boundary component of X has a label in a finite set $L = \{1, a, b, c, \ldots\}$ with involution $\hat{}$, where $\hat{1} = 1$. We call $\hat{a}$ a dual of a. We denote a label set of X for all boundary components by ℓ. Then, let $\mathcal{C} = (X, \ell)$ be a category and $\mathcal{C}'$ be another category such that an object is a finite dimensional complex Hilbert space and a morphism is a unitary transformation.

A functor F from $\mathcal{C}$ to $\mathcal{C}'$ is called a *unitary topological modular functor* if F satisfies the following seven axioms:

(1) (**Disjoint union axiom**): Let X_1 and X_2 be compact oriented surfaces with boundaries whose labels are ℓ_1 and ℓ_2 respectively.

$$F(X_1 \amalg X_2, \ell_1 \amalg \ell_2) = F(X_1, \ell_1) \otimes F(X_2, \ell_2).$$

(2) (**Gluing axiom**): Let $\tilde{X}$ be a compact oriented surface obtained by gluing together a pair of boundary circles with dual labels in X.

$$F(\tilde{X}, \ell) = \bigoplus_{x \in L} F(X, \ell \cup \{x, \hat{x}\}).$$

(3) (**Duality axiom**): Let X^* stnds for X with the reversed orientation and labels applied $\hat{}$.

$$F(X^*) = (F(X))^*, \text{ where we mean that } A^* = {}^t\bar{A}.$$

(4) (**Empty surface axiom**): $F(\emptyset) = \mathbb{C}$.

(5) (**Disk axiom**): Let D be a 2-disk.

$$F(D, \{a\}) = \begin{cases} \mathbb{C} & \text{if } a = 1 \\ 0 & \text{if } a \neq 1 \end{cases}$$

(6) (**Annulus axiom**): Let A be an annulus.

$$F(A, \{a, b\}) = \begin{cases} \mathbb{C} & \text{if } a = \hat{b} \\ 0 & \text{if } a \neq \hat{b} \end{cases}$$

(7) (**Algebraic axiom**): All can be algebraically described over $\mathbb{Q}$ for some bases in $F(D, \{a\})$, $F(A, \{a, \hat{a}\})$ and $F(P, \{a, b, c\})$, where P is a 2-sphere with 3 holes.

Freedman-Larsen-Wang[3] has actually given an example of a unitary topological modular functor by using the $SU(2)$ Chern-Simons-Witten theory at $q = e^{\frac{2\pi i}{5}}$. Note that we have no unitary topological modular functor defined on X with same labels all in the boundaries by the algebraic axiom (7). We also see that a modular functor (see Chapter V in Turaev[9]) defined on X having only label 1's all in the boundary components is nothing but a representation of the braid group.

In physics, there are traditionally laws but no axioms. This is because an old theory has been overtaken by a new theory through clarifying the limit of applicability by a new experimental fact or its self-contradiction. However, it may be useful to clarify the axioms since we have been recently faced with difficulty of experimental inspection. On one hand, in mathematics axiomatization is the standard method.

It will be helpful to recall an axiomatic definition of topological quantum field theory (TQFT) in relation with modular functor. The TQFT in dimension d for a d-dimensional oriented compact smooth manifold Σ is a functor which assigns to a finite dimensional complex vector space $Z(\Sigma)$ and a vector $Z(Y) \in Z(\Sigma)$ such that $\partial Y = \Sigma$. Atiyah[1] has given the axioms in the following:

(1) (**Duality axiom**): $Z(-\Sigma) = Z(\Sigma)^*$, where $Z(\Sigma)^*$ is the dual space of $Z(\Sigma)$.

(2) (**Disjoint union axiom**): $Z(\Sigma_1 \amalg \Sigma_2) = Z(\Sigma_1) \otimes Z(\Sigma_2)$.

(3) (**Associativity axiom**): When $\partial Y_1 = -\Sigma_1 \amalg \Sigma_2$, a linear map $Z(Y_1) : Z(\Sigma_1) \to Z(\Sigma_2)$ is defined and $Z(Y_2) : Z(\Sigma_2) \to Z(\Sigma_3)$ is also defined when $\partial Y_2 = -\Sigma_2 \amalg \Sigma_3$. Then we can construct $Y = Y_1 \cup_{\Sigma_2} Y_2$ by gluing Y_1 and Y_2 along Σ_2.

$$Z(Y) = Z(Y_2) \circ Z(Y_1).$$

(4) (**Empty manifold axiom**): $Z(\emptyset) = \mathbb{C}$.

(5) (**Homotopy invariance axiom**): $Z(\Sigma \times [0, 1]) = \mathrm{id}_{Z(\Sigma)}$.

Note that if Y is a closed $(d+1)$-manifold, then a linear function $Z(Y)$ ia a complex number by Axiom (4), which means that the TQFT has given a topological invariant of a closed $(d+1)$-manifold.

Note also that a vector $Z(Y)$ can be regarded as a linear map. First we recall the well-known fact:

$$V^* \otimes W \cong \text{Hom}(V, W) \tag{3}$$

for two complex vector spaces, where $V^* = \text{Hom}(V, \mathbb{C})$. By combining the duality axiom, disjoint union axiom and (3), for an oriented $(d+1)$ dimensional manifold Y with $\partial Y = (-\Sigma_1) \amalg \Sigma_2$ we have the following:

$$Z(Y) \in Z(\partial Y) = Z((-\Sigma_1) \amalg \Sigma_2) = Z(-\Sigma_1) \otimes Z(\Sigma_2)$$
$$= Z(\Sigma_1)^* \otimes Z(\Sigma_2) \cong \text{Hom}(Z(\Sigma_1), Z(\Sigma_2)).$$

This isomorphism means that a vector $Z(Y)$ is naturally regarded as an element of $\text{Hom}(Z(\Sigma_1), Z(\Sigma_2))$, i.e., a linear map of $Z(\Sigma_1)$ into $Z(\Sigma_2)$).

5. Anyon Model

Anyon is a composite particle which is introduced to explain the phenomena of quantum hall effect, a phenomenon which electron systems confined to a 2-dimensional surface show at low temperature and strong magnetic field. The electrons confined to a metal surface is in superconductive phase under low temperature, and applying strong magnetic field vertically, normal conducting phase appears as holes on the surface. As is well known, the magnetic flux gets driven out in superconductive phase, therefore the flux penetrates into these holes of normal conducting phase, and the system exhibits a state where the flux piercing these holes vertically. Around these holes, the normal conducting electrons (holes) make cyclotron motions. Since these flux penetrating holes on the surface of metal and the normal conducting electrons are correlated, one may expect to treat these as composite particles. That means, we look upon these normal conducting phase holes as a composite particle that are draped with charges and magnetic fluxes as shown in Fig. 4, given the names of Anyons.

When the charged particle rotates around the tube-like magnetic field generated by a solenoid, its wave function ψ deviates by $e^{iq\Phi}\psi$, which is called the Aharonov-Bohm effect. Here, Φ is the total magnetic flux $\Phi = \int_S B dS$ penetrating the tube. Similarly, since anyon is a particle wearing charge and magnetic fluxes, when an electron rotates around the anyon, its phase of wave function also deviates by Aharonov-Bohm effect proportional to the charge. Also, this deviation does not change according to the continuous change of closed paths, i.e. it is a topological invariant.

Consider the probability amplitude $\psi(x_1, x_2)$ of 2 anyons carrying charge q and magnetic flux Φ, to be observed at point x_1 and x_2. As-

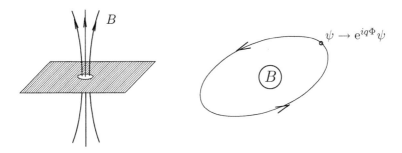

Fig. 4.

sume that the 2nd anyon (at point x_2) went around the first anyon (at point x_1). Then the phase of 2nd anyon's wave function on its first circuit should deviate by $q\Phi$, but since this 2nd particle is also an anyon, it could be interpreted that the 1st particle has rotated around the 2nd anyon. So, the total deviation should be the twice the case an ordinary charge going around a single anyon. Therefore the effect of interchanging the 1st and 2nd anyon would be described as,

$$\psi(x_2, x_1) = e^{i\theta}\psi(x_1, x_2), \quad \theta = q\Phi. \tag{4}$$

The particle which its wave function obeys $\psi(x_2, x_1) = \pm\psi(x_1, x_2)$ are known as bosons or fermions, but as its name shows, the anyon obeys the fractional statistics, $\theta \neq 2\pi, \pi$ in (4). These anyons which obeys such fractional statistics are permitted only on a special 2 dimensional space. When the motion of anyon particles are described on $2 + 1$ dimensional spacetime, one could get a spacetime diagram just like braids, but the anyon system breaks time-reversal symmetry, and can memorize the entanglement of spacetime curves. Which means, one could storage information in this topological links of spacetime curve. The idea of topological quantum computing is to use the anyon system to carryout the quantum computation. Anyon could be thought as a topological soliton solution of the field, and therefore is stable against perturbation. Also the information weaved into the link is a topological value and therefore robust. Creating quantum algorithms could be achieved by braiding the strands. However, while the simple abelian anyon as described above could construct braid group representation, it has only complex 1 dimensional representation by the wave function's phase factor deviation (4). A complete quantum calculation could not be handled by 1 dimensional unitary representation of braid group.

Let aside the existence of nonabelian anyon, shown below is the mathematical model of topological quantum computing by employing nonabelian

anyon.[3,4,10–12]
1) the emerging particles have types and labels
2) rules of fusion and splitting
3) braiding rules when two particles are exchanged
4) braiding rules when two particles are recoupled

(1) (**Labels**): Each particle is labelled by sets: $L = \{1, a, b, c, \ldots\}$. 1 describes no particles, vacuum, and involution $\hat{\ }$ represents charge conjugation. The anti-particle of a is labelled by $\hat{a}$, and for the vacuum $\hat{1} = 1$ is required. The Latin letter a represents irreducible unitary representation, and $\hat{a}$ is its dual representation. From the definition of modular functor given in the previous section, the compact oriented surface X corresponds to the metal plane, the boundaries of X to each anyons, and its label to the types of anyon particles.

(2) (**Fusion rule**): The particle a and b becoming c by fusion is represented by

$$a \times b = \sum_c N_c^{ab} c.$$

Actually it is a composition of unitary representations, for instance, consider 2 adjoint representation 8 of $SU(3)$, then the rule defines that its composition $8 \times 8 = 1 + 8 + 8 + 10 + \hat{10} + 27$ should be taken as $N_8^{88} = 2$. N_c^{ab} is a nonnegative integer. The space of all possible processes of particle a and b to fusion and become a particle c is represented by V_c^{ab} and called a fusion space, and could be made into a vector space by taking an orthonormal basis,

$$\{|ab; c, \mu\rangle, \quad \mu = 1, 2, \ldots, N_c^{ab}\}.$$

$$|ab; c, \mu\rangle = $$
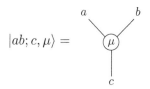

Fig. 5.

The direct product of these vector spaces, $\bigoplus_c V_c^{ab}$ is called the full Hilbert space for anyon pair ab. We assume orthogonality and completeness between these bases,

$$\langle ab; c', \mu' | ab; c, \mu \rangle = \delta_{c'c} \delta_{\mu'\mu}. \quad \sum_{c,\mu} |ab; c, \mu \rangle \langle ab; c, \mu | = I_{ab}.$$

I_{ab} is an identity operator on $\bigoplus_c V_c^{ab}$. The model which satisfies $\sum_c N_c^{ab} \geq 2$ is called the nonabelian anyon model, while those that does not is called the abelian anyon.

(3) **(Braiding rule)**: The swap of the positions of two particles a and b induces an isomorphism mapping $R : V_c^{ab} \to V_c^{ba}$. R is defined by the unitary transformation

$$R|ab; c, \mu \rangle = \sum_{\mu'} |ba; c, \mu' \rangle (R_c^{ab})_{\mu'}^{\mu},$$

and it should be the representation of braid group B_n to satisfy the pentagon and hexagon relation (Sec.6), together with the recoupling map F.

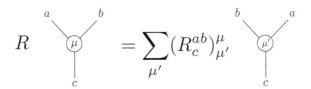

Fig. 6.

(4) **(Recoupling rule)**: There are two natural decomposition for the topological Hilbert space V_d^{abc}, which express three anyons a,b and c fusing to become d. That is,

$$V_d^{abc} \simeq \bigoplus_e V_e^{ab} \otimes V_d^{ec} \simeq \bigoplus_{e'} V_d^{ae'} \otimes V_{e'}^{bc}.$$

Above relation shows that there are two possibilities to be considered, the first a and b fuses and become e, and e and c to be d, and second b and c fuses and become e' and then a and e' fuse to be d. Define the unitary transformation F, an isomorphism of (5) as,

$$F|ab; e, \mu \rangle \otimes |ec; d, \nu \rangle = \sum_{e'\mu'\nu'} |ae'; d, \mu' \rangle \otimes |bc; e', \nu' \rangle (F_d^{abc})_{e'\mu'\nu'}^{e\mu\nu}$$

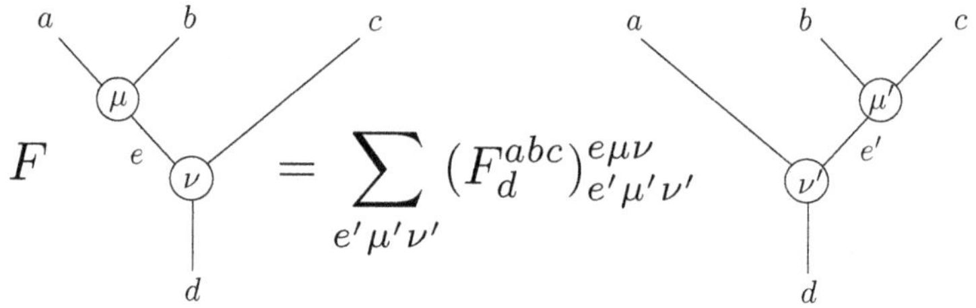

$$F \quad \overset{a \quad b \quad c}{\underset{d}{\overset{\mu}{\underset{\nu}{e}}}} \quad = \sum_{e'\,\mu'\,\nu'} (F_d^{abc})_{e'\mu'\nu'}^{e\mu\nu} \quad \overset{a \quad b \quad c}{\underset{d}{\overset{\mu'}{\underset{\nu'}{e'}}}}$$

Fig. 7.

6. Fibonacci Anyons

In the previous section, we introduced general anyon models. Here, we consider more concrete and simple Fibonacci anyon models.[10,12] In the Fibonacci model there are only two labels, $L = \{1, a\}$, where $\hat{a} = a$. State 1 express vacuum state, and a a state where an anyon particle exist. Further, we require fusion rules between these states,

$$a \times a = 1 + a, \quad a \times 1 = a, \quad 1 \times 1 = 1.$$

Consider topological Hilbert space $V_b^{a_1 a_2 \cdots a_n}$. This represents a fusion space where particles labeled $a_1, a_2, \ldots, a_n$ merges and becomes particle labeled b.

$$|a_1 a_2; b\rangle = \overset{a_1 \qquad a_2}{\underset{b}{\bigvee}}$$

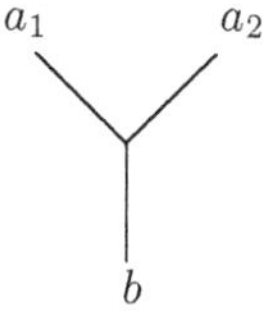

Fig. 8.

There is only one way of two a particles fusing to become one *vacuum*, $V_1^{aa} = < |aa; 1\rangle >_\mathbb{C}$. Writing the fusion space when the numbers of a particles are n as $V_1^{aa \cdots a} = V_1^{(n)}$, $V_1^{(3)} = V_1^{aaa} = < |aa; a\rangle \otimes |aa; 1\rangle >_\mathbb{C}$ and $V_1^{(4)}$ has two possibilities such that could be shown as, therefore,

Fig. 9.

$$V_1^{(4)} = \left\langle \begin{array}{l} |aa;a\rangle \otimes |aa;a\rangle \otimes |aa;1\rangle, \\ |aa;1\rangle \otimes |1a;a\rangle \otimes |aa;1\rangle \end{array} \right\rangle_{\mathbb{C}},$$

similarly we obtain,

$$V_1^{(5)} = \left\langle \begin{array}{l} |aa;a\rangle \otimes |aa;a\rangle \otimes |aa;a\rangle \otimes |aa;1\rangle, \\ |aa;a\rangle \otimes |aa;1\rangle \otimes |1a;a\rangle \otimes |aa;1\rangle, \\ |aa;1\rangle \otimes |1a;a\rangle \otimes |aa;a\rangle \otimes |aa;1\rangle \end{array} \right\rangle_{\mathbb{C}}.$$

Generally, the dimension of $V_1^{(n)}$ is given by,

$$\dim(V_1^{(n)}) = f_{n-2}$$

provided that f_n is Fibonacci's number, $f_{n+2} = f_{n+1} + f_n$, $f_0 = 1$, $f_1 = 1$, which gives this model the name Fibonacci anyons. The recoupling operator F must satisfy the following diagrams as shown in Fig. 10, named as pentagon relations. Writing down these operations as an action to the base

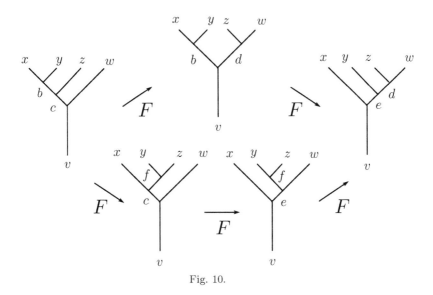

Fig. 10.

vector gives,

$$F_v^{xy(zw)} \circ F_v^{bzw} |((xy; b)z; c)w; v\rangle$$
$$= \sum_{de} |x(y(zw; d); e); v\rangle (F_v^{xyd})_e^b (F_v^{bzw})_d^c$$
$$= F_{yzw}^{yzw} \circ F_v^{x(yz)w} \circ F_c^{xyz} |((xy; b)z; c)w; v\rangle$$
$$= \sum_{def} |x(y(zw; d); e); v\rangle (F_e^{yzw})_d^f (F_v^{xfw})_e^c (F_c^{xyz})_f^b,$$

where we used the notation $|((xy; b)z; c)w; v\rangle = |xy; b\rangle \otimes |bz; c\rangle \otimes |cw; v\rangle$ etc., so the pentagon equation is ,

$$(F_v^{xyd})_e^b (F_v^{bzw})_d^c = \sum_f (F_e^{yzw})_d^f (F_v^{xfw})_e^c (F_c^{xyz})_f^b.$$

Further, the F and braiding operator R should also satisfy the following hexagon relation as shown in Fig. 11. Similarly representing this operation

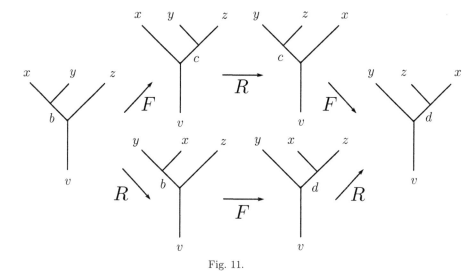

Fig. 11.

for base vectors gives,

$$F_v^{yzx} \circ R_v^{x(yz)} \circ F_v^{xyz} | (xy; b)z; v\rangle$$

$$= \sum_{cd} |y(zx; d); v\rangle (F_v^{yzx})_d^c (R_v^{xc})(F_v^{xyz})_c^b$$

$$= R_{xz}^{xz} \circ F_v^{yxz} \circ R_{xy}^{xy} | (xy; b)z; v\rangle$$

$$= \sum_d |y(zx; d); v\rangle (R_d^{xz})(F_v^{yxz})_d^b (R_b^{xy}),$$

therefore we derive,

$$\sum_c (F_v^{yzx})_d^c (R_v^{xc})(F_v^{xyz})_c^b = (R_d^{xz})(F_v^{yxz})_d^b (R_b^{xy}),$$

the hexagon equation. Especially, for the topological Hilbert space $V_1^{(4)} = V_1^{aaaa}$, the maps $R = R_a^{aa}, R_1^{aa}$ and $F = F_1^{aaaa}, F_1^{1aa}$ has a solution,

$$R_a^{aa} = -e^{\frac{2\pi i}{5}}, \quad R_1^{aa} = e^{\frac{4\pi i}{5}}, \quad (F_a^{aaa})_d^c = \delta_a^c \delta_d^a,$$

$$(F_a^{aaa})_d^c = -\tau \delta_a^c \delta_d^a + \sqrt{\tau} \delta_a^c \delta_d^1 + \sqrt{\tau} \delta_1^c \delta_d^a + \tau \delta_1^c \delta_d^1, \ \tau^2 + \tau = 1.$$

$V_1^{(4)} \simeq \mathbb{C}^2$, and using the basis vectors as shown in Fig. 12, the braiding and recoupling operator, written as a whole $R, F : V_1^{(4)} \to V_1^{(4)}$, becomes

$$R = \begin{pmatrix} e^{\frac{4\pi i}{5}} & 0 \\ 0 & -e^{\frac{2\pi i}{5}}. \end{pmatrix}, \quad F = \begin{pmatrix} \tau & \sqrt{\tau} \\ \sqrt{\tau} & -\tau \end{pmatrix}.$$

Fig. 12.

The generators $\{\sigma_1, \sigma_2\}$ of braid group B_3 on three strands can be represented by

$$\sigma_1 \mapsto R, \quad \sigma_2 \mapsto F^{-1}RF.$$

These matrices generate a representation of the B_3, whose image is dense in $SU(2)$. Similarly in $V_1^{(n)}$, the image of the representation of the braid group is dense in $SU(f_{n-1})$, so we can simulate a quantum computing with Fibonacci anyons by braiding and recoupling.

Appendix A. Fundamental group

A set X (more precisely, (X, d)) is called a metric space if there is a real valued function $d : X \times X \to \mathbb{R}$ such that

(1) $d(x, y) \geq 0$ for any x, $y \in X$; in particular $d(x, y) = 0$ if and only if $x = y$,

(2) $d(x, y) = d(y, x)$ for any x, $y \in X$, and

(3) $d(x, z) \leq d(x, y) + d(y, z)$ for any x, y, $z \in X$.

For example, $(\mathbb{R}^n, d)$ is a typical metric space if we define

$$d(\boldsymbol{x}, \boldsymbol{y}) = \sqrt{(x_1 - y_1)^2 + \cdots + (x_n - y_n)^2}$$

for $\boldsymbol{x} = (x_1, \ldots, x_n)$, $\boldsymbol{y} = (y_1, \ldots, y_n) \in \mathbb{R}^n$. The d is called an Euclidean metric. We have another one; let $C[a, b]$ be the set of all real valued continuous functions defined on the closed interval $[a, b]$. Then we define

$$d(f, g) = \int_a^b |f(x) - g(x)| dx$$

for any f, $g \in C[a, b]$. It is easy to check that $(C[a, b], d)$ is a metric space.

Let (X, d) be a metric space. We define a set

$$B_\varepsilon(a) = \{x \in X; \, d(x, a) < \varepsilon\}$$

for $\varepsilon > 0$ and $a \in X$, which is called an ε-neighborhood of a in X. A subset A is called an *open set* (of X) if it holds that $B_\varepsilon(a) \subset A$ for any $a \in A$ and a well-chosen positive number ε. We say that B is a *closed set* if $X - B$ is an open set of X. It is clear that both the empty set $\emptyset$ and the ambient set X are open sets and also closed sets by definition. Moreover, it is easy to see that $U = \{(x, y) \in \mathbb{R}^2; 0 < y < \frac{\pi}{2}\}$ is an open set of $\mathbb{R}^2$; however, for a continuous map $f : \mathbb{R}^2 \to \mathbb{R}^2$ defined by $f(x, y) = (x \cos y, x \sin y)$ the image $f(U)$ is not an open set of $\mathbb{R}^2$.

Let X be a set and suppose that there is a family of subsets of X, $\mathcal{O}$, such that

(1) $\emptyset \in \mathcal{O}$, $X \in \mathcal{O}$,

(2) if O_1, $O_2 \in \mathcal{O}$, then $O_1 \cap O_2 \in \mathcal{O}$, and

(3) for any family, $\{O_\lambda\}_{\lambda \in \Lambda}$, of subsets in $\mathcal{O}$, $\bigcup_{\lambda \in \Lambda} O_\lambda \in \mathcal{O}$.

Then, X or $(X, \mathcal{O})$ is called a *topological space* and $O \in \mathcal{O}$ an open set of X. Let X_1 and X_2 be topological spaces. We say that a map $f : X_1 \to X_2$ is *continuous* if $f^{-1}(V)$ is an open of X_1 for any open set V of X_2. Note that the image of an open set in the source space is not necessarily open in the target space for a given continuous map, as is already seen above.

Let X be a topological space. For any $p, q \in X$ if there is a continuous map $f : [0,1] \to X$ such that $f(0) = p$, $f(1) = q$, then X is called *arcwise connected* and such a map is called a *path* in X. In what follows, we always consider an arcwise connected topological space and so we refer it simply "a space" hereafter.

Let X be a space, and for p, q, $r \in X$ let f be a path connecting p with q and g a path connecting q with r. Then we define a product of paths, $f \cdot g$, as follows:

$$f \cdot g(s) = \begin{cases} f(2s) & (0 \le s \le 1/2) \\ g(2s - 1) & (1/2 \le s \le 1). \end{cases}$$

The $f \cdot g$ is a new path in X connecting p with r.

When there is a path l such that the starting point and the end point coincide, i.e., a continuous map

$$l : [0,1] \to X, \; l(0) = p, \; l(1) = p$$

is called a *loop* in X with a base point p.

Definition A.1. Let X be a space. For two loops l_0, $l_1 : [0,1] \to X$ with a basepoint p, there is a continuous map $F : [0,1] \times [0,1] \to X$ which satisfies the following conditions:

$$F(t,0) = l_0(t), \; F(t,1) = l_1(t) \quad (\forall t \in [0,1]),$$

then F is called a *homotopy* connecting l_0 with l_1 and we say that l_0 is homotopic to l_1.

Let $L(X,p)$ be the set of all loops with a basepoint p in X. Here we define $l_0 \sim l_1$ if l_0 is homotopic to l_1. Then it is easy to see that

(1) $l_0 \sim l_0$
(2) $l_0 \sim l_1 \implies l_1 \sim l_0$
(3) $l_0 \sim l_1, \; l_1 \sim l_2 \implies l_0 \sim l_2$.

This means that $\sim$ is an equivalence relation in $L(X,p)$. Thus we can consider a quotient space

$$\pi_1(X,p) := L(X,p)/\sim.$$

For any $l \in L(X,p)$, $[l] \in \pi_1(X,p)$ is an equivalence class to which l belongs, which is called a *homotopy class* of l.

Our purpose is to introduce a group structure into this quotient set. If we have done it and it is computable, we can study its geometrical property of X through the group structure. For this purpose we have to define a product in $\pi_1(X, p)$.

First for l, $l' \in L(X, p)$ we define a product of l and l' as follows:

$$l \cdot l' = \begin{cases} l(2t) & (0 \le t \le \frac{1}{2}) \\ l'(2t - 1) & (\frac{1}{2} \le t \le 1). \end{cases}$$

It is easy to see that if for $[l], [l'] \in \pi_1(X, p)$ we take $\tilde{l} \in [l]$, $\tilde{l}' \in [l']$, then we have

$$l \cdot l' \sim \tilde{l} \cdot \tilde{l}'.$$

Thus we reach the following definition:

$$[l] \circ [l'] = [l \cdot l']$$

for $[l]$, $[l'] \in \pi_1(X, p)$, which is well-defined, i.e., does not depend on the choices of representatives.

For $[\alpha], [\beta], [\gamma] \in \pi_1(X, p)$ we have

$$(\alpha \cdot \beta) \cdot \gamma(t) = \begin{cases} \alpha(4t) & (0 \le t \le \frac{1}{4}) \\ \beta(4t - 1) & (\frac{1}{4} \le t \le \frac{1}{2}) \\ \gamma(2t - 1) & (\frac{1}{2} \le t \le 1) \end{cases}$$

and

$$\alpha \cdot (\beta \cdot \gamma)(t) = \begin{cases} \alpha(2t) & (0 \le t \le \frac{1}{2}) \\ \beta(4t - 2) & (\frac{1}{2} \le t \le \frac{3}{4}) \\ \gamma(4t - 3) & (\frac{3}{4} \le t \le 1). \end{cases}$$

Then we can define a homotopy $F : [0, 1] \times [0, 1] \to X$ in the following:

$$F(t, s) = \begin{cases} \alpha(\frac{4t}{s+1}) & (0 \le t \le \frac{s+1}{4}) \\ \beta(4t - s - 1) & (\frac{s+1}{4} \le t \le \frac{s+2}{4}) \\ \gamma(\frac{4t-s-2}{2-s}) & (\frac{s+2}{4} \le t \le 1) \end{cases},$$

which means that $\forall t \in [0, 1]$

$$F(t, 0) = (\alpha \cdot \beta) \cdot \gamma(t), \quad F(t, 1) = \alpha \cdot (\beta \cdot \gamma)(t)$$

and that $\forall s \in [0, 1]$

$$F(0, s) = (\alpha \cdot \beta) \cdot \gamma(0), \quad F(1, s) = \alpha \cdot (\beta \cdot \gamma)(1).$$

Hence we have $(\alpha \cdot \beta) \cdot \gamma \sim \alpha \cdot (\beta \cdot \gamma)$ and the associativity

$$([\alpha] \circ [\beta]) \circ [\gamma] = [\alpha] \circ ([\beta] \circ [\gamma]).$$

Let $e : [0,1] \to X$ be a constant map defined by $e(t) = p$ ($\forall t \in [0,1]$). Then we easily see that for any $[\alpha] \in \pi_1(X, p)$

$$[\alpha] \circ [e] = [e] \circ [\alpha] = [\alpha].$$

Thus $[e]$ is the unit element of $\pi_1(X, p)$. Moreover, for any $[\alpha] \in \pi_1(X, p)$ if we define $\alpha'(t) = \alpha(1 - t)$, then we have

$$[\alpha] \circ [\alpha'] = [\alpha'] \circ [\alpha] = [e].$$

Thus $[\alpha']$ is the inverse element of $[\alpha]$, and hence we have verified that $\pi_1(X, p)$ has a group structure. If we do not need to specify the basepoint $p \in X$, we may omit p and write $\pi_1(X)$ for simplicity because there is an isomorphism $\pi_1(X, p) \to \pi_1(X, q)$ for any p, $q \in X$, where note that this isomorphism is not canonical[c].

For a space X if $\pi_1(X, p)$ is trivial, then we say that X is *simply connected*. For example, an n-dimensional Euclidean space $\mathbb{R}^n$, an n-dimensional disk $D^n = \{(x_1, \ldots, x_n) \in \mathbb{R}^n; x_1^2 + \cdots + x_n^2 \leq 1\}$, $\mathbb{R}^n - \{(0, \ldots, 0)\}$ with $n \geq 3$, and $(n - 1)$-dimensional sphere $\partial D^n = S^{n-1}$ with $n \geq 3$ are all simply connected. On one hand, a circle S^1, $\mathbb{R}^2 - \{(0, \ldots, 0)\}$, and $SO(n)$ with $n \neq 1$ are not simply connected.

Let X and Y be spaces and $f : X \to Y$ be a continuous map such that $f(x_0) = y_0$. Take a loop $l : [0,1] \to X$ with a basepoint x_0 and consider the composite map

$$f \circ l : [0,1] \xrightarrow{\;l\;} X \xrightarrow{\;f\;} Y,$$

which is also a loop in Y with a basepoint y_0. If a loop l_1 in X is homotopic to l_2, then $f \circ l_1$ is also homotopic to $f \circ l_2$.

Namely, given a continuous map $f : X \to Y$, we can obtain a map by considering the correspondence $[l] \in \pi_1(X, x_0)$ to $[f \circ l] \in \pi_1(Y, y_0)$. Then we write this correspondence as follows:

$$f_* : \pi_1(X, x_0) \longrightarrow \pi_1(Y, y_0)$$
$$\rotatebox{90}{$\in$} \qquad\qquad\qquad \rotatebox{90}{$\in$}$$
$$[l] \qquad \longmapsto \qquad [f \circ l]$$

---

[c]If we take a path connecting p with q, then this isomorphism is defined. Namely, the isomorpohism does depend on the choice of the path.

We can easily check that f_* is a homomorphism. f_* is called the induced homomorphism by f.

Furthermore, let Z be a space and $g : Y \to Z$ be a continuous map. Then, the composite map

$$X \xrightarrow{f} Y \xrightarrow{g} Z$$

induces the homomorphism

$$(g \circ f)_* : \pi_1(X, x_0) \to \pi_1(Z, z_0)$$

and we immediately see that $(g \circ f)_* = g_* \circ f_*$.

Let $f : X \to Y$ be a homeomorphism such that $f(x_0) = y_0$. Then, since both f and f^{-1} are continuous, we have

$$f_* \circ (f^{-1})_* = (f \circ f^{-1})_* = (\mathrm{id}_Y)_* = \mathrm{id}_{\pi_1(Y)}$$
$$(f^{-1})_* \circ f_* = (f^{-1} \circ f)_* = (\mathrm{id}_X)_* = \mathrm{id}_{\pi_1(X)}.$$

This means that f_* is an isomorphism (see Exercise 2.1 (5)). Thus we have obtained

Theorem A.1. *Let X and Y be arcwise connected topological spaces. If X is homemorphic to Y, then $\pi_1(X)$ is isomorphic to $\pi_1(Y)$. In other words, if $\pi_1(X)$ is not isomorphic to $\pi_1(Y)$, then X is not homeomorphic to Y.*

If X is not homeomorphic to Y, then by definition there is no homeomorphism between them. However, it is extremely difficult to show directly that no homeomorphism exists. We emphasize here that only the difference of algebraic structures of fundamental groups immediately provides homeomorphism criterion.

Moreover, the theorem has a slight generalization. Let X and Y be spaces. If there are continuous maps $f : X \to Y$ and $g : Y \to X$ such that $g \circ f$ is homotopic to id_X and $f \circ g$ is homotopic to id_Y, then we say that X is homotopy equivalent to Y. As is easily seen, we can replace "homeomorphic" by "homotopy equivalent" in Theorem A.1.

Further, if X is homotopy equivalent to a point, then we say that X is *contractible*. Note that if X and Y are homeomorphic, then they are homotopy equivalent but the converse implication does not hold in general, e.g. both $\mathbb{R}$ and $\mathbb{R}^2$ are contractible (and hence homotopy equivalent) but not homeomorphic. It is easy to show that $\mathbb{R}$ is not homeomorphic to $\mathbb{R}^2$. Suppose that they are homeomorphic. Then, there is a homeomorphism

$f : \mathbb{R} \to \mathbb{R}^2$. Since f is an injection, there is $x \in \mathbb{R}$ such that $f(x) = (0,0)$. Consider the restricted map $f|_{\mathbb{R}-\{x\}} : \mathbb{R} - \{x\} \to \mathbb{R}^2 - \{(0,0)\}$, which is still a homeomorphism. Then, $\mathbb{R} - \{x\}$ is not connected but $\mathbb{R}^2 - \{(0,0)\}$ is connected, which contradicts the fact that the map $f|_{\mathbb{R}-\{x\}}$ is a homeomorphism since the connectedness is topological invariant. This can be deduced from the calculations that $\pi_1(\mathbb{R} - \{x\}) = \{e\}$ and $\pi_1(\mathbb{R}^2 - \{(0,0)\}) \cong \mathbb{Z}$ (see Exercise 6.1 and Exercise 6.2 below). In general, the invariance of dimension of an Euclidean space holds, i.e., $\mathbb{R}^m$ is homeomorphic to $\mathbb{R}^n$ if and only if $m = n$, which can be proved by using the notion of "homology group" but we omit it here because we need more pages in order to discuss homology.

Exercise 6.1. Show that there is an isomorphism $\pi_1(S^1, x_0) \cong \mathbb{Z}$ according to the following orders: First we set $S^1 = \{(x,y) \in \mathbb{R}^2; x^2 + y^2 = 1\}$ and take a basepoint $x_0 = (1,0)$. Moreover, consider a continuous map $\varphi : \mathbb{R} \to S^1$ defined by $\varphi(x) = (\cos x, \sin x)$.

(1) Let $f : [0,1] \to S^1$ be a continuous map such that $f(0) = x_0$. Then there uniquely exists a map $\tilde{f} : [0,1] \to \mathbb{R}$ such that

$$\tilde{f}(0) = 0, \quad \varphi \circ \tilde{f} = f.$$

(2) Define a map $\psi : \pi_1(S^1, x_0) \to \mathbb{Z}$ by $\psi([f]) = \tilde{f}(1)$. Show that

(a) ψ is a homomorphism.
(b) ψ is a bijection.

Exercise 6.2.

(1) We have a canonical isomorphism

$$\pi_1(X \times Y, (x_0, y_0)) \cong \pi_1(X, x_0) \times \pi_1(Y, y_0).$$

(2) Show that a map $f : \mathbb{R}^2 - \{(0,0)\} \to S^1 \times \mathbb{R}$ defined by

$$f(x,y) = \left(\frac{x}{\sqrt{x^2 + y^2}}, \frac{y}{\sqrt{x^2 + y^2}}, \log(x^2 + y^2) \right)$$

is a diffeomorphism.

(3) Show that $\mathbb{R}^2$ is not homotopy equivalent to $\mathbb{R}^2 - \{(0,0)\}$ by calculating their fundamental groups.

Van Kampen's Theorem

We would like to carry out its calculation of fundamental group for a topological space. As is easily expected, a topological space with a simple geometrical structure has its fundamental group with a simple group structure. However, fundamental group can be generally non-abelian by definition. Thus we need to know better how to find group in order to calculate fundamental group effectively for a given topological space. In this section we further introduce Van Kampen's theorem which is useful in calculation of fundamental group.

Let X and Y be arcwise connected topological spaces, $X \cap Y$ be also. Then, Van Kampen's theorem tells us a relation among those fundamental groups of

$$\pi_1(X \cup Y), \ \pi_1(X), \ \pi_1(Y), \ \pi_1(X \cap Y).$$

Theorem A.2 (Van Kampen). *(1) For arcwise connected topological spaces X, Y, $X \cap Y$, we take a base point $x_0 \in X \cap Y$. If $\pi_1(X \cap Y, x_0) \cong \{e\}$, then we have*

$$\pi_1(X \cup Y, x_0) \cong \pi_1(X, x_0) * \pi_1(Y, x_0).$$

(2) Let X be a connected polyhedron and $X \cup D^2$ be a space attaching the 2-disk along the boundary circle. We take a basepoint $x_0 \in X \cap D^2 = \partial D^2 = S^1$. Let ℓ be a homotopy class in $\pi_1(X, x_0)$ represented by a loop $X \cap D^2$ and (ℓ) denote the normal subgroup of $\pi_1(X, x_0)$ generated by ℓ. Then we have

$$\pi_1(X \cup D^2, x_0) \cong \pi_1(X, x_0)/(\ell)$$

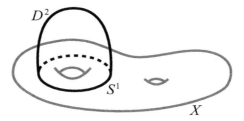

Fig. A1.

We omit the proof (see e.g. Hatcher[5]). We should emphasize here that it is more important to master how to use Van Kampen's theorem than to understand its proof.

Take two circles and attach at one point x_0 (see the figure below). This

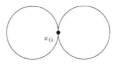

Fig. A2. A bouquet $S^1 \vee S^1$

space is called a bouquet of S^1, denoted by $S^1 \vee S^1$.

By using Theorem A.2 (1) we have $\pi_1(S^1 \vee S^1, x_0) \cong \mathbb{Z} * \mathbb{Z}$. Let Int D^2 be a space homeomorphic to a set $\{(x, y) \in \mathbb{R}^2; x^2 + y^2 < 1\}$. Then, we see that $\pi_1(S^1 \times S^1 - \text{Int } D^2) \cong \mathbb{Z} * \mathbb{Z}$ since $S^1 \times S^1 - \text{Int } D^2$ is homotopy equivalent to $S^1 \vee S^1$. If we take generators α and β in $\pi_1(S^1 \times S^1 - \text{Int } D^2)$, then we see that the boundary loop ∂D^2 represents $\alpha^{-1}\beta^{-1}\alpha\beta$. Thus, by Theorem A.2 (2) together with Exercise 2.2, we have

$$\pi_1(S^1 \times S^1) \cong \mathbb{Z} * \mathbb{Z}/(\alpha^{-1}\beta^{-1}\alpha\beta) = \langle \alpha, \beta \,|\, \alpha^{-1}\beta^{-1}\alpha\beta = e \rangle \cong \mathbb{Z} \oplus \mathbb{Z},$$

since $S^1 \times S^1 = (S^1 \times S^1 - \text{Int } D^2) \cup D^2$.

We have another typical example. Let us calculate the fundamental group of the real projective plane $\mathbb{R}P^2 = M \cup D^2$ by using Theorem A.2, where M is the Möebius band. First note that M is homotopy equivalent to S^1, and hence $\pi_1(M) \cong \mathbb{Z}$. Take a generator $\alpha \in \pi_1(M)$ represented by a central circle of the band M (see the dotted line in Figure A3). Then, we immediately see that the boundary loop ∂D^2 is α^2. Thus, by Theorem A.2, we have

$$\pi_1(\mathbb{R}P^2) \cong \mathbb{Z}/(\alpha^2) = \langle \alpha \,|\, \alpha^2 = e \rangle \cong \mathbb{Z}_2.$$

We will leave several handy exercises:

Exercise 6.3.

(1) Find the fundamental group of a closed orientable surface Σ_g with genus $g \geq 2$.

88

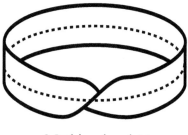

Möbius band M

Fig. A3.

(2) Show that the fundamental group of the Klein bottle is

$$\langle \alpha, \beta \mid \alpha^2 \beta^2 = e \rangle$$

and its abelianization, i.e.,

$$\langle \alpha, \beta \mid \alpha^2 \beta^2 = e, \ \alpha^{-1} \beta^{-1} \alpha \beta = e \rangle$$

is isomorphic to $\mathbb{Z} \oplus \mathbb{Z}_2$.

(3) Show that $\pi_1(SO(3)) \cong \mathbb{Z}_2$ and $\pi_1(U(2)) \cong \mathbb{Z}$.

References

1. M. F. Atiyah, *Topological Quantum Field Theories*, Publ. Math. IHES **68** (1989) 175–186.
2. J. S. Birman, *Braids, links and mapping class groups*, Ann. of Math. Studies, vol. 82 (Princeton Univ. Press, 1974).
3. M. H. Freedman, M. Larsen and Z. Wang, *A modular functor which is universal for quantum computation*, Comm. Math. Phys. **227** (2002) 605–622.
4. M. H. Freedman, A. Kitaev, M. Larsen and Z. Wang, *Topological Quantum Computation*, quant-ph/0101025.
5. A. Hatcher, *Algebraic Topology* (Cambridge Univ. Press, 2002).
6. V. F. Jones, *A polynomial invariant for knots via von Neumann algeras*, Bull. Amer. Math. Soc. **12** (1985) 103–111.
7. M. Nakahara, *Overview on Quantum Computing*, Lectures in summer school at Kinki University, August 2007.
8. V. V. Prasolov and A. B. Sossinsky, *Knots, Links, Braids and 3-manifolds*, Translations of Mathematical Monographs, vol. 154 (Amer. Math. Soc., 1997).
9. V. G. Turaev, *Quantum invariants of knots and 3-manifolds*, de Gruyter Studies in Math., vol. 18 (Berlin-New York, 1994).
10. J. Preskill, *Lecture Notes for Physics 219: Quantum Computation*, http://www.iqi.caltech.edu/preskill/ph219.

11. A. Marzuoli and M. Rasetti, *Computing spin neworks*, Ann. Physics **318** (2005) 345–407.
12. L. H. Kauffman and S. J. Lomonaco Jr., *q-deformed spin networks, knot polynomials and anyonic topological quantum computation*, J. Knot Theory Ramifications **16** (2007) 267–332.

AN INTRODUCTION TO ENTANGLEMENT THEORY

DAMIAN J. H. MARKHAM

Department of Physics, Graduate School of Science, University of Tokyo, Tokyo 113-0033, Japan

We introduce the theory of entanglement intended for an audience of physicists, computer scientists and mathematicians not necessarily having a background in quantum mechanics. We cover the main concepts of entanglement theory such as separability, entanglement witnesses, LOCC and entanglement measures. Along the way we will see many interesting questions arise spanning mathematics, physics and information science amongst other disciplines.

Keywords: Entanglement Theory, Separability, LOCC.

1. Introduction

Entanglement arises naturally from the mathematical formalism of quantum mechanics. From a physical perspective this property is however pretty surprising and has led to many longstanding debates amongst physicists as to the validity and meaning of this formalism. More recently, entanglement has become seen as a resource for quantum information processing, useful for absolutely secure key distribution and quantum computation.

The theory of entanglement developed over the past two decades shows rich mathematical structure, with many subtleties (both from physics and a mathematics perspectives) and has drawn ideas and techniques from such diverse fields as thermodynamics, convex analysis, information theory and group theory. Despite the vast progress made over the years there are still an exciting number of questions remaining in many directions, in its use in quantum information, in the mathematical challenges it presents and in its fundamental role as a part of physics.

The intention of this introduction is to take the student through the main concepts of entanglement theory as clearly as possible. It will not and cannot be a definitive guide to the subject (for a pretty good attempt at this please see the Horodeckis' review[1]) and of course some of the topics covered will be influenced by my own field of research to an extent. However I will endeavor to cover all the basics and point out interesting open questions when they arise, relations to other fields and further reading. I hope that these notes will pass on some of my own enthusiasm for what I think is an incredibly exciting field to work in.

Mathematically entanglement is simply a consequence of composite systems in quantum mechanics. This will be our starting point in these lectures. We will begin in section 2 by introducing quantum states, statespace and how operations on states are represented mathematically. Moving to composite systems we will see entanglement emerge as a property of quantum mechanics in section 3. We will then move to the quantification of entanglement in section 4. First we explain the general setting of LOCC, then move to how this is used to define different measures of entanglement. We will also briefly discuss methods of measuring entanglement and some subtleties which arise in the multipartite case.

2. Quantum Mechanics and State Space

In order to facilitate a speedy review of the formalism of quantum mechanics we will simply present the most general way to express states and evolution of quantum systems, without too much recourse to the physical motives behind them. We hope that this will also be more amenable to mathematicians and computer scientists to get the important points. From a physical perspective this route may seem a little topsy turvy, but hopefully this is countered by its simplicity and transparency. For a more detailed account describing many of the physical motivations for the axioms taken, please see Ref. 2.

2.1. *State space*

In quantum mechanics, to all degrees of freedom of physical objects, we associate a Hilbert state $\mathcal{H}$. For example an electron is described using $\mathcal{L}_2$ for the position degree of freedom, and $\mathbb{C}_2$ for the spin degree of freedom. A particular system at any point in time is then described by a "state". The state describes the physical properties of the system - that is, what would happen if we were to measure the system. Mathematically, a state is described by a *density matrix* ρ living in the operator space over the Hilbert space $\mathcal{H}$ of the system at hand. The set of density matrices are the set of all operators satisfying

$$\rho^\dagger = \rho$$
$$\mathbf{1} \geq \rho \geq 0$$
$$\mathrm{Tr}[\rho] = 1. \tag{1}$$

That is, any operator acting over Hilbert space $\mathcal{H}$ and satisfying the above properties can be said to describe the state of a physical system. We refer to the space of all density operators as *state space* $\mathcal{S}(\mathcal{H})$. It can easily be checked that state space is convex - i.e. convex combinations of density matrices are also density matrices.

All density matrices can be diagonalised to their eigenbasis

$$\rho = \sum_i p_i |e_i\rangle\langle e_i|, \tag{2}$$

where the $|e_i\rangle$ are the eigenstates (elements of the underlying Hilbert space $\mathcal{H}$), and the p_i can be interpreted as probabilities since by definition we have $1 \geq p_i \geq 1$ and $\sum_i p_i = 1$. If the sum in (2) only consists of one element $\rho = |\psi\rangle\langle\psi|$, we say it is a *pure state* (in most introductory undergraduate courses in quantum mechanics we only consider pure states). In such a case we often simply write the unique eigenstate $|\psi\rangle$. Thus one interpretation of the density matrix is of a classical mixture, or ensemble, of the pure states $\{p_i, |e_i\rangle\}$. A state which is not pure is called *mixed*. A simple way to test if a given density matrix is pure or mixed is the fact that ρ is pure if and only if $\rho^2 = \rho$ (as can be easily checked).

In fact any density matrix has many possible decompositions into convex combinations of pure states. Thus, the *ensemble* interpretation is not unique - physically this means that many different physical situations may give rise to the same physical state. For example, take two decompositions of some state

$$\rho = \sum_i p_i |\psi_i\rangle\langle\psi_i|$$
$$= \sum_j q_j |\phi_j\rangle\langle\phi_j|, \tag{3}$$

this could be created by randomly sampling the states $\{|\psi_i\rangle\}$ according to the probability distribution p_i, or by randomly sampling the states $\{|\phi_j\rangle\}$ according to q_j. The freedom in decomposition occurs again and again in entanglement theory and is the cause of some great difficulty when it comes to checking the existence of entanglement, as well as its quantification.

Note that one can in fact take a more abstract mathematical approach in terms of C* algebras, which can become important when considering infinities, but for the situation covered in these lectures (and for most of quantum information) this reduces to the formalism described above. From here on, unless stated otherwise we are considering finite dimensional systems, where the above completely suffices.

In quantum information we usually do not specify the physical system at hand. Instead we take a general two level system known as a *qubit* (and leave the physics of implementation to someone else!). A general pure qubit is written

$$|\psi\rangle = \alpha|0\rangle + \beta|1\rangle, \qquad (4)$$

with $|\alpha|^2 + |\beta|^2 = 1$. Its density matrix is given by

$$\rho = |\psi\rangle\langle\psi| = \begin{pmatrix} |\alpha|^2 & \alpha\beta^* \\ \alpha^*\beta & |\beta|^2 \end{pmatrix}, \qquad (5)$$

written in the eigenbasis (i.e. $|\psi\rangle$) and the *computational basis* $\{|0\rangle, |1\rangle\}$.

2.2. *Evolution*

All evolutions of a quantum system can be mathematically expressed as completely positive (CP) maps from one space of density matrices $\mathcal{S}_1$ to another $\mathcal{S}_2$ (in fact we will usually consider $\mathcal{S}_1 = \mathcal{S}_2$). Completely positive maps are extremely important in the theory of entanglement as we will see in later sections. For now we skip the definition of complete positivity until section 2.4, and will instead give the most general way to express such maps. These fall into two classes of operations, those that do not preserve the trace 1 property of density matrices (or rather those which require an additional normalisation element to the formalism), and those that do - completely positive trace preserving (CPTP) maps.

CP maps not preserving the trace property can be thought of as resulting from an evolution where at some point a measurement has taken place, who's outcome m is known

$$\varepsilon^m(\rho) = \sum_i E_i^m \rho E_i^{m\dagger}, \qquad (6)$$

such that $\sum_i E_i^{m\dagger} E_i^m \leq 1$. The E_i^m are known as the *Kraus operators*. The superscript m can be seen as indicating the classical information of some measurement outcome in the evolution process. The trace of $\varepsilon^m(\rho)$ may be less than one (if $\sum_i E_i^{m\dagger} E_i^m < 1$), which is why we say these maps are not trace preserving. The full normalised version (necessary for the output to correspond to a state) is given by

$$\rho \rightarrow \frac{\varepsilon^m(\rho)}{\text{Tr}[\varepsilon^m(\rho)]}. \tag{7}$$

The class of CPTP maps can be understood as having resulted from an evolution where the results m are erased or forgotten, hence we must average over all m. Then we have

$$\rho \rightarrow \varepsilon(\rho) = \sum_m \varepsilon^m(\rho)$$
$$= \sum_{m,i} E_i^m \rho E_i^{m\dagger}, \tag{8}$$

with $\sum_{i,m} E_i^{m\dagger} E_i^m = \mathbf{1}$. Note that CPTP maps are obviously a subclass of CP maps, so all CPTP maps can of course be written in the form of equation (6), by simple relabelling. So in fact (6) is the most general form of evolution possible. Hence in some sense index m acts as a dummy index and, although the interpretation of m as a measurement outcome is very useful (and often physically accurate), we must be a little bit careful as it may not always be the case.

When there is only one term in the summation (6), and if the trace remains 1, the evolution is *unitary*[a]. This is a clear example of a CPTP map which does not involve a measurement. But this seems a bit cheating as there is only one term. An important property of unitary operations is that it is impossible to go from a pure state to a mixed state, which can be easily checked. We can see unitary maps simply as a change of basis of the Hilbert space. In general measurement does cause mixing (by adding extra terms), but we must be careful to note that it is not the only way to generate mixing. It is possible to have CPTP mixing with many terms not via measurement as we will see in later sections - via entanglement.

2.3. POVMs, projective measurement and observables

We have seen above the most general evolution of a quantum state. Let us now go back and extract the essential features of the measurement part. (Usually in text books you will find these introduced in the reverse order and in fact the idea of CP maps can be derived from axioms about measurements

---

[a]An operator U is unitary iff $U^\dagger U = \mathbf{1}$.

e.g. Ref. 2). Sometimes it is not necessary to know the full evolution of the system when considering a measurement, say, for state estimation, we may just need to know the statistics of the outcomes of the measurements (rather than also knowing the resultant state as well). This leads to the idea of the formalism of POVMs.

To any measurement on a state ρ we associate a set of operators $\{M_m\}$, collectively known as the *Positive Operator Valued Measure (POVM)* which obey

$$\sum_m M_m = \mathbb{1}$$
$$\mathbb{1} \geq M_m \geq 0 \;\; \forall m, \tag{9}$$

where m represents an outcome of the measurement, and the probability of outcome m occurring p_m is given by $p_m = \text{Tr}[\rho M_m]$.

There is no mention of how the state actually evolves, and indeed for any given POVM, there is no unique way to implement it in general. This can be seen if we write the POVM in terms of general evolution operators. The measurement would evolve the system to an ensemble as

$$\rho \rightarrow \{p_m, \sum_i E_i^m \rho E_i^{m\dagger}\},$$
$$M_m = \sum_i E_i^{m\dagger} E_i^m. \tag{10}$$

Clearly in general there are many different Kraus operators E_i^m which would give the same M_m, so the actual physical process is not unique in general. Thus POVMs represent only the statistics of a measurement.

Relating back to the map (6) we see that the ensemble is written $\{p_m, \sum_i \varepsilon^m(\rho)\}$ and the m represents the outcomes as mentioned. Forgetting the outcomes would then be the same as taking the ensemble to the mixed state $\sum_m p_m \varepsilon^m(\rho)$.

Any measurement can be written as a POVM, and in turn any set of operators satisfying properties (9) represents the POVM of a physically valid measurement.

If we look at the action of a unitary rotation of a state $\rho \rightarrow U\rho U^\dagger$, we see that the statistics of any measurement would be changed $p_m \rightarrow \text{Tr}[U\rho U^\dagger M_m] = \text{Tr}[\rho U^\dagger M_m U]$ by the cyclic property of the trace operator. Thus we can consider a unitary rotation of a state ρ as simply a rotation of the basis in which we measure (this is the essence of the Heisenberg representation of quantum evolution).

An important class of POVMs that we will come across again are *projective measurements*. These are measurements whose POVM operators P_m satisfy

$$P_m^2 = P_m. \tag{11}$$

They can always be expanded as $P_m = \sum_\alpha |\alpha\rangle\langle\alpha|$ for some orthogonal basis $|\alpha\rangle$. By comparing with (9) we see that in this case the Kraus operators are fixed, hence we have both the statistics and the outcome states of the measurement.

To complete the review of measurements we introduce observables. An *observable* is a Hermitian operator, written in the eigenbasis

$$A = \sum_m a_m P_m, \tag{12}$$

where the eigenvalues a_m represent the outcome of some experiment, and the P_m are projections onto the eigenspaces. These are the normal representation of measurements in an undergraduate course on quantum mechanics. Examples of such observables which are common in quantum information are the Pauli spin operators. The expectation value for a given state ρ (i.e. the average outcome) is given by

$$\langle A \rangle_\rho := \sum_m p_m a_m = \mathrm{Tr}[A\rho] \tag{13}$$

2.4. Composite systems

To combine more than one quantum system we employ the tensor product formalism. For systems with associated Hilbert spaces $\mathcal{H}_A$ and $\mathcal{H}_B$, the states are density matrices over $\mathcal{H}_A \otimes \mathcal{H}_B$.

For a composite state $\rho_{A,B}$, if we restrict ourselves to one subsystem, $\mathcal{H}_A$, the *reduced state* is given by the

$$\rho_A := \mathrm{Tr}_B[\rho_{A,B}], \tag{14}$$

where Tr_B is the *partial trace* and is defined as

$$\mathrm{Tr}_B[\rho_{A,B}] := \sum_\alpha \langle\alpha|_B \rho_{A,B} |\alpha\rangle_B, \tag{15}$$

where $\{|\alpha\rangle_B\}$ form an orthonormal basis over $\mathcal{H}_B$. This definition of reduced density matrix is equivalent to defining it as the operator ρ_A which satisfies $\mathrm{Tr}[\rho_A X] = \mathrm{Tr}[\rho_{A,B} X \otimes \mathbf{1}]$ for all X (see e.g. Ref. 3). Operationally this basically says that it is the only way to represent a reduced state such that measurements on one system faithfully can be seen as measurements on half of the global system.

As a side note, with the introduction of composite systems, a simple and elegant picture of general quantum operations emerges. It can be shown (see e.g. Ref. 2) that all quantum evolutions, which take the form (7) can be regarded as combinations of adding an ancillary system, performing unitary operations, measuring of the ancilla system and tracing over the ancilla system. We will sometimes find this picture of evolution helpful for descriptive reasons, but mathematically we can always use (6,8).

We are now in a position to define complete positivity. A positive map Θ is *completely positive* iff

$$\Theta \otimes \mathbf{1} \geq 0 \qquad (16)$$

for any finite dimensional identity operator $\mathbf{1}$. That is, if and only if the positivity remains with arbitrary extension. This concept is massively important in entanglement theory as we will see in later sections. An important example of an operator that is positive, but not completely positive is the partial transpose which we will meet in section 3.

Finally we now come to entanglement. Mathematically it arises from the observation that not all states on a composite of N systems $\mathcal{H} = \mathcal{H}_A \otimes \mathcal{H}_B \otimes ... \otimes \mathcal{H}_N$ can be written in the form

$$\rho = \sum_i p_i \rho_A^i \otimes \rho_B^i \otimes ... \otimes \rho_N^i. \qquad (17)$$

States that can be written in this form are called *separable*, and states that cannot be written in this form are called *entangled*.

3. Entanglement and Separability

Now we have a definition of what entanglement is, let us look at the simple example of a bipartite state, at first considering pure states. This will allow us to introduce many terms in a simple way. Following our description of quantum states for the pure state case, the most general pure state in a system composed of two systems A and B is written as $|\psi\rangle_{AB} \in \mathcal{H}_A \otimes \mathcal{H}_B$,

$$|\psi\rangle_{AB} = \sum_{i,j} a_{i,j} |i\rangle_A \otimes |j\rangle_B \equiv \sum_{i,j} a_{i,j} |ij\rangle_{AB}, \qquad (18)$$

where $\{|i\rangle_A\}$ and $\{|j\rangle_B\}$ are orthonormal bases for $\mathcal{H}_A$ and $\mathcal{H}_B$ respectively, the tensor product of which is often shortened to $|i\rangle_A \otimes |j\rangle_B \equiv |ij\rangle_{AB}$, and $a_{i,j}$ are complex coefficients satisfying $\sum_{i,j} a_{i,j} = 1$.

The definition of separability for pure states reduces to the set of states which can be written in the form

$$|\psi\rangle_{AB} = |a\rangle_A \otimes |b\rangle_B, \tag{19}$$

for some states $|a\rangle \in \mathcal{H}_A$ and $|b\rangle \in \mathcal{H}_B$. Separable pure states are often referred to as *product states*.

An extreme example of an entangled pure state in $\mathbb{C}_2 \otimes \mathbb{C}_2$ is the following

$$|\Phi^+\rangle_{AB} := \frac{1}{\sqrt{2}}(|00\rangle_{AB} + |11\rangle_{AB}). \tag{20}$$

It can easily be seen that there is no way of decomposing this state into a product state as in (19) by contradiction. This state is a *maximally entangled* state of $\mathbb{C}_2 \otimes \mathbb{C}_2$ [b].

If we look at the reduced state of either subsystem, we see how entanglement implies some local mixedness. The reduced state density matrices are

$$\rho_A = \rho_B = \begin{pmatrix} 1/2 & 0 \\ 0 & 1/2 \end{pmatrix}. \tag{21}$$

This is already in its eigenbasis, so it is clearly not a pure state. This is the *maximally mixed* state of $\mathbb{C}_2 \otimes \mathbb{C}_2$ [c]. In fact, the amount of mixedness is a good measure of how entangled a pure state is (as we will see in later sections).

In the bipartite pure state case there is a very convenient way to write states (as far as their entanglement is concerned). For all states $|\psi\rangle_{AB} \in \mathcal{H}_A \otimes \mathcal{H}_B$, there exist local basis $|i\rangle_A \in \mathcal{H}_A$ and $|i\rangle_B \in \mathcal{H}_B$, such that the state can be written

$$|\psi\rangle_{AB} = \sum_{i=1}^{r} \sqrt{p_i}|ii\rangle_{AB}. \tag{22}$$

This is known as the *Schmidt decomposition*, the coefficients p_i are the *Schmidt coefficients* and the minimum r is known as the *Schmidt number* (known to be a good quantifier of entanglement as we will also see in later sections). This decomposition captures all the features of the entanglement of the system, so is very important in entanglement theory.

---

[b] A maximally entangled state of $\mathbb{C}_d \otimes \mathbb{C}_{d'}$, $d \le d'$ is given by $|\Phi^+\rangle_{AB} = \sum_{i=0}^{d-1} \frac{1}{\sqrt{d}}|i, i\rangle_{AB}$.
[c] A maximally mixed state of $\mathbb{C}_d$ is given by $\rho = \mathbf{1}/d$.

From the the symmetry of the Schmidt decomposition, for pure bipartite states we can see that we have

$$\rho_A = \rho_B. \tag{23}$$

Also a pure state is entangled iff the Schmidt decomposition has only one term (i.e. the Schmidt number is equal to one). Thus, for a global pure bipartite state, the state is entangled iff the reduced states are mixed.

Now let us turn to mixed states. The first thing we note is that the last observation above no longer holds - reduced states can be mixed even without the presence of entanglement. Indeed, as we will see more in subsequent sections, the pure bipartite state case is rather special in the ease with which we can analyse entanglement theory.

Our task in testing separability is to test if a state can be written in the form

$$\rho_{AB} = \sum_i p_i \rho_A^i \otimes \rho_B^i. \tag{24}$$

For the bipartite mixed state case a useful tool can be found looking at the action of certain maps on the state. In fact the completely positive maps which occurred earlier. Consider a positive map acting on one part of the system Λ_B, which may not be completely positive. Clearly for all separable states we have

$$(\mathbf{1} \otimes \Lambda_B(\rho_{AB})) \geq 0, \tag{25}$$

since $\Lambda_B(\rho_B^i) = \rho_B^{i'}$ is a valid density matrix, thus taking any separable state to another one just priming the right hand part of (24).

In fact it has been shown[4] that a bipartite state ρ_{AB} is separable iff for all positive maps $\Lambda : \mathcal{S}(\mathcal{H}_B) \to \mathcal{S}(\mathcal{H}_A)$ the condition $(\mathbf{1} \otimes \Lambda(\rho_{AB})) \geq 0$ holds.

Thus in the bipartite case, the question of separability becomes one of checking this condition over all positive maps which are not completely positive.[d]

In the case of $\mathcal{C}_2 \otimes \mathcal{C}_2$ and $\mathcal{C}_2 \otimes \mathcal{C}_3$, the map of partial transposition is such a strong example of a positive positive, but not completely positive map as

---

[d]Note that positive, but not completely positive maps are not actual physical processes. The question then arrises: can such tests be carried out physically? Seemingly not. However one approach to this has been to study approximate maps, which while CP, approximate well the non-CP maps above. See for example Ref. 5.

to cover all possibilities. For systems residing over $\mathbb{C}_2 \otimes \mathbb{C}_2$ or $\mathbb{C}_3 \otimes \mathbb{C}_2$, all positive maps $\Lambda : \mathbb{C}_d \to \mathbb{C}_{d'}$, $d = 2$, $d' = 2, 3$, can be decomposed as

$$\Lambda = \Lambda_{\mathrm{CP}}^{(1)} + \Lambda_{\mathrm{CP}}^{(2)} \circ T. \tag{26}$$

From this it follows that taking transposition over one system is sufficient to check for separability here.

Restated, for state $\rho_{\mathrm{AB}} = \sum_{i,j,k,l} a_{i,j,k,l} |ij\rangle\langle kl|$, the *partial transposition* over subsystem $\mathcal{H}_{\mathrm{B}}$ is given by

$$\rho_{\mathrm{AB}}^{T_{\mathrm{B}}} = \sum_{i,j,k,l} a_{i,l,k,j} |ij\rangle\langle kl|, \tag{27}$$

and is dependant on the basis $|ij\rangle$ at hand. Then the above statement is written: A state $\rho_{\mathrm{AB}} \in \mathbb{C}_2 \otimes \mathbb{C}_2$ or $\rho_{\mathrm{AB}} \in \mathbb{C}_3 \otimes \mathbb{C}_2$ is separable iff its partial transpose is positive $\rho_{\mathrm{AB}}^{T_{\mathrm{B}}} > 0$.

This is known as the PPT condition for separability, initially presented by Peres[6] as a necessary condition for separability and later used for the above theorem. The fact that it is not in general a sufficient condition was shown first in Ref. 7 where an example of an entangled PPT state was presented.

When it comes to higher dimensions and larger number of systems, the problem of being able to check if a state is separable or not becomes extremely difficult, and in fact poses one of the largest open problems in the theory of entanglement.

The task is to find out if a given state can be written in a decomposition in the form of (17). Since, as mentioned, there are in general infinitely many possible decompositions for a state, this is a difficult task to check and we should need some clever mathematical tricks. In the bipartite case this can be seen as the problem of describing positive but not completely positive maps in a concise useful way (a recognised difficult problem in mathematics). In the multipartite case it is even more difficult.

Several approaches have been developed which work well for different sets of states. Broadly speaking there are two sets of results, those necessary conditions (such as PPT) which are implied by separability (but do not necessarily imply it), and those that are necessary and sufficient, which are generally difficult to use or of limited application. It is beyond the scope of these lectures to go into details and we refer the reader to the Horodeckis review[1] for a good overview.

Let us now take a small step back and review what the set of separable states looks like. We define

$$\text{SEP} := \{\rho_{\text{SEP}} | \exists \text{ decomposition } \rho_{\text{SEP}} = \sum_i p_i \rho_A^i \otimes \rho_B^i \otimes ... \otimes \rho_N^i\}. \quad (28)$$

We can see that this set is also convex - a probabilistic mixture of different separable states is also separable. Also, it can be shown that the volume of separable states is non-zero around the identity[8] for any finite dimensional system.

The convexity of this set actually becomes incredibly useful and allows for a method of separating sets of entangled states from the separable ones, via entanglement witnesses.[4,9] These are operators which define hyperplanes separating some set of entangled states from the convex set of separable states. Not only are these operators interesting from a mathematical point of view, they are also Hermitian and so represent physical observables. Thus their role in entanglement theory spreads to the experimental validation of entanglement too.

The basis behind these witnesses is the Hahn-Banach theorem which states that given a convex body and a point x_o that does not belong to it, then there exists a linear function f that takes positive values for all points belonging to the convex body, while $f(x_0) < 0$ (see Ref. 10 for example).

Stated in terms of density operators we define an *entanglement witness* W as Hermitian operators satisfying

$$W \geq 0$$
$$\forall \rho_{\text{SEP}} \in \text{SEP}, \ \text{Tr}[W\rho_{\text{SEP}}] \geq 0$$
$$\exists \rho \notin \text{SEP} \text{ s.t. } \text{Tr}[W\rho] < 0 \quad (29)$$

The Hahn Banach theorem then states that for all entangled ρ there exists a witness which can see it.

In fact these witnesses are related to the separation of separable and entangled states given by the positive operators mentioned above (see e.g. Ref. 1,10). From a physics perspective, they are also related to Bell's inequalities - in fact Bell's inequalities are examples of entanglement witnesses which also witness the violation of local realism.

Entanglement witnesses are incredibly important in experimental verification of entanglement. An important question that arises there is finding the optimal decomposition into local observables. That is, in situations concerning entanglement we are almost by definition separate, so we would like to measure these witnesses locally. We would then like to decompose these into sets of local measurements in as efficient a way as possible. In certain cases optimality can be shown, and this is an important ongoing line of investigation.

The practical usefulness of entanglement witnesses also extends to the quantification of entanglement. In certain cases the amount of violation of the entanglement witness (that is, how negative is the outcome) gives a bound to the 'amount' of entanglement present[11–13] (a concept which we will expand on in the next section).

This simple and elegant approach of entanglement witnesses to the study of entanglement is especially valuable when considering large complicated systems such as those in condensed matter,[14,15] where often we cannot even be sure of the state of the system exactly and typically measurements accessible are averaged over many subsystems (in the thermodynamic limit infinitely many). A process which often irons out any quantum effects. Incredibly it is shown that even such 'rough' measurements as the magnetic susceptibility can act as witnesses of entanglement.[14]

4. Quantification of Entanglement

In the last section we have seen the definition of the term entanglement, and how, given a state we may decide whether it is entangled or not. Now we would like to quantify how entangled a state is. For example, consider the state

$$\sqrt{1 - \delta}|00\rangle + \sqrt{\delta}|11\rangle. \tag{30}$$

This state is clearly entangled when $\delta = 1/2$ and separable when $\delta = 0$ or $\delta = 1$. Its symmetry indicates that $\delta = 1/2$ may be somehow the most entangled, and so what about the in between bits?

Before we can really answer this question we must define better the problem setting. Entanglement is a property of quantum states which is inherently non-local, and has meaning when considering systems which are separated. This idea is made precise in the paradigm of local operations and classical communication.

4.1. *Local operations and classical communication*

Quantification of entanglement is studied and defined with respect to a special set of operations encapsulating the idea of what we mean by entanglement. These are *Local Operations and Classical Communication* (LOCC).

The situation is one of separate laboratories. We consider a system spread over several subsystems, where each subsystem is acted upon separately. For a system on $\mathcal{H}_A \otimes \mathcal{H}_B \otimes ... \mathcal{H}_N$, LOCC maps are those comprised

of combinations of
- Local CPTP maps on the individual subsystems
- Classical Communication

The role of the classical communication is to allow the generation of classical correlations. The motivation behind this paradigm is essentially to test which properties of the system persist beyond local quantum effects, allowing for global "classical" effects - that is correlations generated by classical communication (for a fuller discussion on the physical motivation for these conditions, see e.g. Ref. 1,16).

It is apparent that all separable states can be made by LOCC. To see this we only need to look at the definition of a separable state (17). Taking the ensemble interpretation, this can be generated for example, by the experimentalist at system A randomly choosing a letter i according to probability distribution p_i [e], then sending the choice to all the experimentalists at the remaining systems, according to which i they all locally generate the state ρ_x^i. Then if all experimentalists erase all memory of which i they had, the overall state is that described by (17). Similarly states which can be generated by LOCC are separable. Thus we have an interpretation of separable states as those which can be created by LOCC and entangled states as those which cannot.

Let us now try to write down the Kraus operators corresponding to an LOCC operation. We first consider the bipartite case. We now slip into the standard notation of calling one system $\mathcal{H}_A$, belonging to Alice, and the other $\mathcal{H}_B$, to Bob. Without loss of generality we assume that Alice goes first. She may perform some measurement and send the result to Bob, so then his action would depend on Alice's outcome. Then, supposing he does a measurement too, Bob could send Alice his result and she could make an operation according to that, e.t.c. In this way the Kraus operators would become correlated, and the map would be of the form

$$\Lambda_{\text{LOCC}}(\rho) = \sum_{a_1,b_1,a_2..} ...(E_A^{a_1,b_1,a_2} \otimes \mathbf{1})(\mathbf{1} \otimes E_B^{a_1,b_1})(E_A^{a_1} \otimes \mathbf{1})\rho$$
$$(E_A^{a_1} \otimes \mathbf{1})^\dagger(\mathbf{1} \otimes E_B^{a_1,b_1})^\dagger(E_A^{a_1,b_1,a_2} \otimes \mathbf{1})^\dagger... \quad (31)$$

(similar holds for the multipartite case). Given the possibility of going on and on with the correlations, the sum can go on for arbitrarily many terms.

---

[e] If this sounds too non-physics we can always imagine the i as outputs of a quantum measurement.

Indeed the difficulty in writing LOCC operations in a concise way is one of the most difficult obstacles to overcome in entanglement theory.

For this reason we often consider simpler sets of operations, such as *separable operations* (those mapping separable states to separable states), or *PPT operations* (those mapping PPT states to PPT states). These operations have much simpler mathematical form, and so have allowed many results to be found where restricting to the stricter set of LOCC operations is too complicated. For now however we will stick to considering LOCC.

4.2. Entanglement measures

Now we have the situation clearly established (i.e. LOCC), we can begin to ask what this setting would have to say about quantifying the amount of entanglement present in some given state ρ. We can roughly split results into two main approaches to this question; an operational approach, where entanglement is defined with respect to how useful this state would be to some quantum information task; and an abstract approach, where we define an entanglement measure as some functional on the set of states which obeys certain axioms.

So far in these lectures we have taken a rather abstract approach to entanglement theory, but this is really not how the field began. People became interested in entanglement in the early 90s when it was noticed that it could be useful for certain tasks [f]. This approach led to the first attempts to quantify entanglement. The archetypal state used in quantum information is the bipartite state which we already know as the maximally entangled state

$$|\Phi^+\rangle = \frac{1}{\sqrt{2}}(|00\rangle + |11\rangle). \qquad (32)$$

This is the "most" useful state of two qubits in several respects: it (or local unitary equivalent states) gives the optimum fidelity in teleportation (i.e. perfect teleportation),[17] it also allows for perfect secrecy and efficiency in quantum key distribution.[18,19] On the other hand taking a state as in (30) with $\delta \neq 1/2$ these tasks cannot be done as well. Indeed the ability to do these tasks depends on the value of δ in a very precise way. This is not a particular feature of these tasks or this size of system, in general we will see

---

[f]Although of course entanglement has been an issue in physics since at least 1935, entanglement theory proper started only then. A kind of market driven start to a research field.

that the entanglement properties are well parameterised by the Schmidt coefficients of bipartite pure states.

With this in mind, the idea of the usefulness of a quantum state was developed viewing the maximally entangled (ME) state as an ideal resource state. The question asked was, given several copies of a quantum state shared between Alice and Bob, how many ME pairs can they get by LOCC, and conversely, how many ME pairs does it take to generate a given state. Using well developed ideas of asymptotics from information theory the entanglement quantities of *distillable entanglement* E_D and *entanglement cost* E_C were presented respectively.[20]

More precisely, we define the distillable entanglement as the optimum rate for distilling maximally entangled states in the limit of infinite number of copies, optimized over all possible LOCC protocols.

$$E_D := sup\{r : \lim_{n \to \infty} (\inf_{\Lambda_{LOCC}} ||\Lambda_{LOCC}(\rho^{\otimes n}) - |\Phi_{2^{rn}}^+\rangle||) = 0\}, \quad (33)$$

where $|\Phi_{2^{rn}}^+\rangle$ is the ME state on $\mathcal{C}_d \otimes \mathcal{C}_d$ with dimension $d = 2^{rn}$ (which is equivalent to rn ME pairs of qubits), $||.||$ is the trace norm (see e.g. Ref. 21) and Λ_{LOCC} is an LOCC map. Similarly the entanglement cost is defined as the asymptotic rate going from ME pairs to copies of the state at hand.

$$E_C := sup\{r : \lim_{n \to \infty} (\inf_{\Lambda_{LOCC}} ||\rho^{\otimes n} - \Lambda_{LOCC}(|\Phi_{2^{rn}}^+\rangle)||) = 0\}. \quad (34)$$

There are other operational measures of entanglement, for example the fidelity (a measure of the faithfulness of the process) of teleportation itself can be considered as an entanglement measure. For now however, we will leave the operational approach and go back to axiomatic thinking.

Axiomatically, we say that any entanglement measure is a map $E :$ $S(\mathcal{H}_A \otimes \mathcal{H}_B \otimes ... \otimes \mathcal{H}_N) \to \mathbb{R}^+$ such that it also obeys:
1) Monotonicity under LOCC maps Λ_{LOCC}

$$E(\Lambda_{LOCC}(\rho)) \leq E(\rho). \quad (35)$$

More precisely we ask that it decreases on average. That is for an

$$E(\rho) \geq \sum_m p_m E\left(\varepsilon_{LOCC}^m(\rho)\right), \quad (36)$$

where $\Lambda_{LOCC}(\rho) = \sum_m \varepsilon_{LOCC}^m(\rho)$ takes the state to ensemble $\{p_m, \varepsilon_{LOCC}^m(\rho)\}$ where m is the (known) outcome of some measurement.
2) Vanishing on separable states, $E(\rho_{SEP}) = 0$ for all $\rho_{SEP} \in SEP$.

We call any such functional obeying these properties an *entanglement monotone*.

This list of axioms is somehow a minimum requirement of a good entanglement measure and some authors also include additional axioms, e.g. of *additivity, continuity* and *convexity*. Additivity and continuity are important when it comes to considering infinite limits (of dimension and of number of copies in entanglement manipulation) and convexity sometimes added because of its mathematical convenience.

Notice that monotonicity under LOCC implies that local unitary (LU) invariance of entanglement monotones. We can see this since unitaries are reversible. If an LU operation U decreases entanglement, then $U^\dagger$, which is also LU can increase entanglement causing a contradiction. This of course makes sense since LU operations are just local changes in the basis and should not effect global properties[g].

An example of a measure satisfying all of the above, often used in quantum information is the *relative entropy of entanglement*[22]

$$E_R(\rho) = \min_{\omega \in \text{SEP}} D(\rho||\omega), \tag{37}$$

where $D(\rho||\omega) = \text{Tr}[\rho(\log_2 \rho - \log_2 \omega)]$ is the relative entropy. This is an example of a so-called "distance-like" measure of entanglement, where we look for the closest separable state with respect to some distance-like function (in this case the relative entropy). Of course we must be careful with which function to minimise over to check that it is an entanglement monotone.

Of course there are many more entanglement measures in exsistence, but this lecture course is too short to cover them all and rather we refer to larger reviews.[1,16]

4.3. Uniqueness of measures, order on states

A question which naturally arises is, is there a unique measure of entanglement within this framework. The simple answer is no (at least if we just talk about entanglement monotones). This can be seen in the bipartite case by the seminal result of Nielsen,[23] which states the following. Given two bipartite states $|\psi\rangle$ and $|\phi\rangle$, with ordered Schmidt coefficients $\{\psi_i\}$ and $\{\phi_i\}$, it is possible to convert $|\psi\rangle \rightarrow |\phi\rangle$ by LOCC perfectly (with unit probability)

---

[g]Note that global basis change can of course generate (or degrade) entanglement. This may seem obvious but sometimes it can cause confusion to physicists in trying to decide if a system can be said to be 'entangled' and there could be two or more basis physically accessible, but entanglement exists only in one. We must be careful in physics to state clearly which global basis we are talking about.

iff $\{\psi_i\}$ are *majorized* by $\{\phi_i\}$, that is iff for all $k > 0$ we have

$$\sum_{i=1}^{k} \psi_i \leq \sum_{i=1}^{k} \phi_i. \tag{38}$$

Since majorisation is a partial order, we see that LOCC induces only a partial order on the set of bipartite pure states. Thus there exist pairs of states such that

$$|\psi\rangle \nleftrightarrow_{\text{LOCC}} |\phi\rangle. \tag{39}$$

It is therefore possible to define entanglement monotones such that

$$E_1(|\psi\rangle) > E_1(|\phi\rangle)$$
$$E_2(|\psi\rangle) < E_2(|\phi\rangle). \tag{40}$$

However, for bipartite pure states at least, this problem can be smoothed over when considering asymptotics. It can be shown[24] that by including certain axioms, that for pure states all measures coincide to the *entropy of entanglement*[20]

$$E_{ent}(|\psi\rangle_{\text{AB}}) := S(\rho_{\text{A}}) = -\sum_{i} p_i \log_2 p_i, \tag{41}$$

where ρ_{A} is the reduced state of subsystem A (or B equivalently), $S(\rho) = -\text{Tr}[\rho \log_2 \rho]$ is the von Neumann entropy (which is a measure of the mixedness of a state) and p_i are the Schmidt coefficients of $|\psi\rangle_{\text{AB}}$.

In fact, under these additional axioms, it can be shown that for bipartite mixed states all such measures are bounded by E_{D} (Eqn. (33)) and E_{C} (Eqn. (34)) (Ref. 24)

$$E_{\text{D}} \leq E \leq E_{\text{C}}. \tag{42}$$

One may ask, then, if there is a better way of defining entanglement such that there is always a unique measure (not just for the bipartite pure state case). For example, could this be possible by adding new axioms or changing those we have? However, such approaches have been studied a lot in the literature and so far no good answer has been found.

So it seems for now at least that we are stuck with having to accept that there are different *types* of entanglement, corresponding to different measures. So we can ask, what does that mean? May it be that different types of entanglement are useful for different quantum information tasks? Or could they hold some different significance in different areas of physics? Interestingly we see then that even in our attempt to find a clean axiomatic mathematical approach, we are in the end drawn back to questions of the operational value of our theory of entanglement.

4.4. *Measuring entanglement*

Coming now back to physics for a bit, we would like to be able to know what we can say about viewing the presence of entanglement in actual experiments.

The most direct approach to this is total state tomography. This is where we repeatedly prepare and measure a state to build up statistics up to the point when we finally get all elements of the density matrix. From this we can use the mathematical tools developed so far to answer questions of separability and amount of entanglement. However this approach is experimentally very costly and we would prefer a more efficient strategy.

One approach we heard about before in section 3 concerning treating non-CPTP maps experimentally. In Ref. 25 and developed in Ref. 5 (amongst others) an ingenious general approach is given to approximating experimentally any non-linear function on states by making global measurements on several copies. Such techniques can be used to check separability and approximate the value of entanglement measures.

An even weaker (but still incredibly powerful) approach we have also discussed before is that of entanglement witnesses. As mentioned, not only can entanglement witnesses be used to separate entangled states from the set of separable states, but also in certain cases their violation can be used to bound entanglement[11,12,26] and as an entanglement measure itself.[26] The main idea is, given the expectation value of a set of witnesses $\{W_k\}$ we want to find the worst case corresponding value for entanglement. For some chosen entanglement measure E, this gives the bound

$$\epsilon(\omega_1, ..., \omega_n) := \inf\{E(\rho)|\mathrm{Tr}[W_k\rho] = \omega_k \forall k\}. \tag{43}$$

This can be phrased in terms of Legendre transformations, and has been calculated for several witnesses and entanglement measures and further applied to experiments, amazingly allowing some old results for the calculation of a witness to give bounds on the value of entanglement.

4.5. *Multipartite entanglement*

We finish our review of entanglement with a discussion about the multipartite case. It is no coincidence that most of the clean results we have presented are for the bipartite case (if not just the bipartite pure). We have seen the difficulty in deciphering the structure of statespaces induced by entanglement in the bipartite case. It gets even worse for the multipartite case of three or more systems.

Even under a simpler set of operations of stochastic LOCC (SLOCC) - LOCC allowing for non unit probability of success (CP LOCC maps with known outcome) - we have inequivalent states for three or more parties.[27] For four or more there are possibly even infinitely many different classes of states under SLOCC. Thus we have many different 'types' or entanglement.

We are then forced to take steps to simplify the situation. One approach is to consider which kinds of states are 'normal'. That is, which kinds of states are likely to occur in any given situation. This can lead to a much simpler induced structure of state space (e.g. Ref. 28). However many questions still remain as to how to define well the notion of 'likely'. Typically we take the approach of random sampling, but we would really like some physical reason to exclude certain states (initial answers are proposed in Ref. 29). Indeed, even if some states are 'rare' physically, it may be that they are so useful to us that we would like to still use them in quantum information.

Another approach that bares fruit is to take simple measures of entanglement such as distance-like measures. The simplicity of the definition skips much of the detailed structure of the state space. Although the defining optimization makes calculation difficult, in some cases (including many states useful in quantum information) symmetries can be used to calculate entanglement and entanglement properties.

Although this class of measures misses out on much of the richness of the structure of state space under LOCC, this approach does allow some freedom in choosing which 'type' of entanglement is considered. As stated in (37), it is completely unclear where the entanglement arises in these kinds of measures. For example a GHZ state

$$|\text{GHZ}\rangle_{\text{ABC}} = \frac{1}{\sqrt{2}}(|000\rangle_{\text{ABC}} + |111\rangle_{\text{ABC}}), \qquad (44)$$

has the same relative entropy of entanglement as

$$|\Phi\rangle_{\text{ABC}} = \frac{1}{\sqrt{2}}(|00\rangle_{\text{AB}} + |11\rangle_{\text{AB}}) \otimes |\phi\rangle_{\text{C}}, \qquad (45)$$

where $|\phi\rangle_{\text{C}}$ is an arbitrary single qubit state, with $E_{\text{R}}(|\text{GHZ}\rangle) = E_{\text{R}}(|\Phi\rangle) = 1$, even though the entanglement is clearly different. By choosing E_{R} with respect to different cuts, we can discriminate these states. Calling $E_{\text{R}}^{\text{Bi(AB:C)}}$ the relative entropy of entanglement with respect to the set of separable states across the bipartite cut AB : C, we have $E_{\text{R}}^{\text{Bi(AB:C)}}|\text{GHZ}\rangle = 1$ and $E_{\text{R}}^{\text{Bi(AB:C)}}|\Phi\rangle = 0$. Hence by analysing different cuts we can get some structure from entanglement also.

Of course, the simplest approach is that of entanglement witnesses mentioned before. This is particularly successful in the multipartite case since other approaches are so difficult. It can also be adapted to consider different types of entanglement. Again by being careful in specifying which separable set we compare a given state to, we can witness whether a state is entangled over all systems, or just over some subsystems.

5. Conclusions

In these notes we have attempted to present the basics of entanglement theory. We have taken a rather mathematical approach, and left the physical motivations to further reading. Having said that we have seen interesting issues raised about the physical measurement of entanglement and mentioned a few possible approaches.

Of course, we have only skimmed the surface of this fertile field, and so to finish, it is perhaps fitting to list a few of the important topics we have *not* covered in these notes (where direct references are not given, references can be found in Ref. 1).

- Infinite dimensional systems. In all of the results here we have only considered finite sized systems. As hinted at, when going to infinities, there are many more subtleties that must be addressed, such as continuity. Indeed entanglement may not be continuous in state space for this case (though some continuity can be regained by certain 'physical' restrictions such as finite energy).[30]
- Connection to quantum channels. There are a plethora of results relating the ability to generate/manipulate entanglement to the quality of a quantum channel as an information carrier. This approach has fielded many useful results in both directions, including bounds on how good channels are in terms of entanglement and new information based entanglement measures such as the *squashed entanglement*.[31] Also, through equivalence of problems it has yielded one of the largest remaining problems of entanglement theory (next point).
- Additivity. That is given several copies of a state, how does entanglement scale (roughly speaking). Through connections to communication theory mentioned above, the question of additivity of the entanglement of formation is shown to have consequences in the ability for additivity of communication quantities such as the Holevo bound.
- Use of entanglement in quantum information. We have only briefly touched upon the usefulness of entanglement as a resource here. There

are in fact many open questions remaining about its use in quantum information, particularly in the multipartite case. In quantum communication our understanding of the role of entanglement is pretty restricted to the bipartite case; could it be that the subtleties of multipartite entanglement have much more to offer, perhaps different classes of entanglement useful for different tasks? In quantum computation we are only making the first steps to understand exactly the role and requirement of entanglement (in particular, we have recent progress in measurement based quantum computing[32]).

- Role of entanglement in many-body physics. This question is developing into one of the most active areas of current entanglement research. See e.g. Ref. 33 for a recent review. Many interesting questions arise from both physical and mathematical points of view as to the relationship between entanglement theory and the theory of criticality in many body physics.

- The physics of entanglement. In these notes we have taken largely an abstract approach to the whole subject. In real physical systems there are other considerations that must be taken into account. Firstly, as mentioned briefly, in any system, the choice of how we split our space is a physical one, and so the definition of LOCC must depend on the physics at hand. Secondly there are often many more constraints in a physical system as to the allowed set of operations and states, for example those induced by symmetries. It is another ongoing field of research to try to impose these extra fundamental physical laws to our theory of entanglement.

References

1. R. Horodecki, P. Horodecki, M. Horodecki and K. Horodecki, quant-ph/0702225.
2. M. A. Nielsen and I. L. Chuang, *Quantum Computation and Quantum Information* (Cambridge University Press, 2000).
3. M. Hayashi, *Quantum Information An Introduction* (Springer-Verlag, 2006).
4. M. Horodecki, P. Horodecki and R. Horodecki, *Phys. Lett. A* **223** (1996) 1.
5. H. Carteret, *Phys. Rev. Lett.* **94** (2005) 040502.
6. A. Peres, *Phys. Lett. A* **128** (1988) 19.
7. P. Horodecki, *Phys. Lett. A* **232** (1997) 333.
8. K. Życzowski, P. Horodecki, A. Sanpera and M. Lewenstein, *Phys. Rev. A* **58** (1998) 883.
9. B. M. Terhal, *Phys. Lett. A* **271** (2000) 319.
10. I. Bengtsson and K. Życzkowski, *Geometry of Quantum States* (Cambridge University Press, 2006).

11. O. Gühne, M. Reimpell and R. Werner, *Phys. Rev. Lett.* **98** (2007) 110502.
12. J. Eisert, F. B. ao and K. Audenaert, *New J. Phys.* **9** (2007) 46.
13. K. M. R. Audenaert and M. B. Plenio, *New J. Phys.* **8** (2006) 266.
14. Č. Brukner and V. Vedral quant-ph/0406040.
15. G. Tóth, *Phys. Rev. A* **71** (2005) 010301.
16. M. Plenio and S. Virmani, *Quant. Inf. Comp.* **7** (2007) 1.
17. C. H. Bennett, G. Brassard, C. Crepeau, R. Jozsa, A. Peres and W. K. Wootters, *Phys. Rev. Lett* **70** (1993) 1895.
18. C. H. Bennett and G. Brassard (1984), in Proc. IEEE International Conference on Computers, Systems, and Signal Processing, Bangalore, India (IEEE, New York, 1984).
19. A. Ekert, *Phys. Rev. Lett.* **67** (1991) 661.
20. C. H. Bennett, H. J. Bernstein, S. Popescu and B. Schumacher, *Phys. Rev. A* **53** (1996) 2046.
21. E. Rains, *Phys. Rev. A* **60** (1999) 173.
22. V. Vedral and M. B. Plenio, *Phys. Rev. A* **57** (1998) 1619.
23. M. A. Nielsen, *Phys. Rev. Lett.* **83** (1999) 436.
24. M. Donald, M. Horodecki and O. Rudolph, *J. Math. Phys.* **43** (2002) 4252.
25. A. Ekert, C. M. Alves, D. K. L. Oi, M. Horodecki, P. Horodecki and L. Kwek, *Phys. Rev. Lett.* **88** (2002) 217901.
26. F. G. S. L. Brandão, *Phys. Rev. A* **72** (2005) 022310.
27. W. Dür, G. Vidal and J. I. Cirac, *Phys. Rev. A* **62** (2000) 062314.
28. P. Hayden, D. Leung and A. Winter, *Comm. Math. Phys.* **265** (2005) 95.
29. O. Dahlsten, R. Oliveira and M. B. Plenio, *J. Phys. A: Math. Theo.* **40** (2007) 8081.
30. J. Eisert, C. Simon and M. B. Plenio, *J. Phys. A* **35** (2002) 3911.
31. M. Christandl and A. Winter, *J. Math. Phys.* **45** (2004) 829.
32. M. V. den Nest, W. Dür, A. Miyake and H. J. Briegel, *New J. Phys.* **9** (2007) 204.
33. L. Amico, R. Fazio, A. Osterloh and V. Vedral, quant-ph/0703044.

HOLONOMIC QUANTUM COMPUTING AND ITS OPTIMIZATION

SHOGO TANIMURA

Department of Applied Mathematics and Physics, Graduate School of Informatics,
Kyoto University, Kyoto 606-8501, Japan
E-mail: tanimura@i.kyoto-u.ac.jp

Holonomic quantum computer uses a non-Abelian adiabatic phase associated with a control-parameter loop as a unitary gate. Holonomic computation should be processed slowly for keeping adiabaticity. On the other hand, holonomic computation should be executed in a short time for avoiding decoherence. These two requests, the slow process and the quick execution, seem contradictory each other and lead to a kind of optimization problem, which is called the isoholonomic problem. In this lecture we review the formulation of the holonomy in quantum mechanics and the solution of the optimal control problem of the holonomy. We also discuss possible implementations of the holonomic quantum computer.

Keywords: Quantum Computation, Berry Phase, Holonomy, Optimization, Isoholonomic Problem, Horizontal Extremal Curve.

1. Introduction

Many designs of practical quantum computers have been proposed in this decade. Holonomic quantum computer is one of those candidates and has been studied mostly in the theoretical aspect. A few years ago Hayashi, Nakahara and I obtained an exact result about the optimal control problem of the holonomic quantum computer.[1,2] In this lecture I aim to explain the basic of holonomic quantum computation. I will give a quick review on Berry phase and Wilczek-Zee holonomy and introduce the idea of holonomic quantum computation. I will formulate the optimization problem of holonomic quantum computer and show its solution. I will apply the result to the two well-known unitary gates; the Hadamard gate and the discrete Fourier transformation gate. Finally, I will briefly discuss the remaining problems, which concern implementation of holonomic quantum computation in physical systems. Most of this lecture note depends on our papers.[1,2]

2. Holonomies in Mathematics and Physics

2.1. *Holonomy in Riemannian geometry*

The first example of holonomy can be found in geometry.[3,4] When a tangent vector is parallel transported along a closed loop in a Riemannian manifold, the vector gets a net rotation. The net rotation generated by the parallel transportation is called the holonomy. In this case, the holonomy manifests curvature of the manifold. The concept of parallel transportation has been generalized in various contexts and described in terms of the connection or the gauge field as discussed below.

2.2. *Berry phase in quantum mechanics*

We shall review the adiabatic theorem[5,6] in quantum mechanics and the formulation of Berry phase.[7] Comprehensive reviews on this subject have been published.[8–10] Let us consider a quantum system whose Hamiltonian $H(\lambda)$ depends on a set of parameters $\lambda = (\lambda^1, \ldots, \lambda^m)$. In practice λ might be an electric field or a magnetic field applied on the system. These parameters can be controlled externally and are specified as functions of time t. The whole of possible values of λ forms a parameter space M. Suppose that the Hamiltonian has a discrete spectrum

$$H(\lambda)\,|n;\lambda\rangle = \varepsilon_n(\lambda)\,|n;\lambda\rangle \tag{1}$$

for each value of λ. The state $|\psi(t)\rangle$ evolves obeying the Schrödinger equation

$$i\hbar\frac{d}{dt}|\psi(t)\rangle = H(\lambda(t))\,|\psi(t)\rangle. \tag{2}$$

Suppose that at $t = 0$ the initial state is an eigenstate of the initial Hamiltonian as $|\psi(0)\rangle = |n;\lambda(0)\rangle$. Assume that no degeneracy occurs along the course of the parameter $\lambda(t)$. Namely, it is assumed that $\varepsilon_n(\lambda(t)) \neq \varepsilon_{n'}(\lambda(t))$ for any $n'(\neq n)$ and any t. Furthermore, let us assume that we change the control parameter $\lambda(t)$ as slowly as possible. The slowness will be defined at (7) in more rigorous words. Then the *adiabatic theorem*[5,6] tells that the state $|\psi(t)\rangle$ remains an eigenstate of the instantaneous Hamiltonian $H(\lambda(t))$ at each time t:

$$|\psi(t)\rangle = e^{i\gamma(t)}|n;\lambda(t)\rangle. \tag{3}$$

In other words, if the external parameter changes slowly enough, and if no level crossing occurs, then transition between different energy levels does not occur and the quantum state remains quasi-stationary in some sense.

Suppose that at a time T the parameter $\lambda(t)$ returns to the initial value as $\lambda(T) = \lambda(0)$. Then the state returns to the initial state up to a phase factor:

$$|\psi(T)\rangle = e^{i\gamma(T)}|\psi(0)\rangle. \tag{4}$$

The phase γ is given as a sum $\gamma = \gamma_d + \gamma_g$ with

$$\gamma_d = -\frac{1}{\hbar}\int_0^T \varepsilon_n(\lambda(t))\,dt, \tag{5}$$

$$\gamma_g = i\int_0^T \langle n;\lambda(t)|\frac{d}{dt}|n;\lambda(t)\rangle\,dt = i\int_C \sum_{\mu=1}^m \langle n;\lambda|\frac{\partial}{\partial\lambda^\mu}|n;\lambda\rangle\,d\lambda^\mu. \tag{6}$$

A proof of these formulae will be given around (16) in a more general context. The phase γ_d is called the *dynamical phase*. The phase γ_g is called the *geometric phase* or the *Berry phase*. It is independent of the time-parametrization of the curve C in the control parameter space.

Sufficient slowness of movement of the control parameter is characterized by the condition

$$\{\varepsilon_n(\lambda(t)) - \varepsilon_{n'}(\lambda(t))\}T \gg 2\pi\hbar \tag{7}$$

for $n' \neq n$ during $0 \leq t \leq T$. This inequality is called the adiabaticity condition. When two distinct energy levels approach closely or cross, the adiabaticity condition is violated.

2.3. *Wilczek-Zee holonomy in quantum mechanics*

Immediately after Berry formulated his celebrated phase, which belongs to the one-dimensional unitary group $U(1)$, Wilczek and Zee[11] formulated its generalization to an arbitrary unitary group $U(k)$. Here we shall describe their formulation. Suppose that the Hamiltonian has a discrete spectrum but now degenerate eigenvalues:

$$H(\lambda)|n,\alpha;\lambda\rangle = \varepsilon_n(\lambda)|n,\alpha;\lambda\rangle. \tag{8}$$

Here $\alpha = 1, 2, \ldots, k_n$ is an index to label degenerate energy eigenstates. It is assumed that these eigenvectors are normalized as $\langle n,\alpha;\lambda|n',\alpha';\lambda\rangle = \delta_{nn'}\delta_{\alpha\alpha'}$. It is said that the energy eigenvalue $\varepsilon_n(\lambda)$ is k_n-fold degenerate. Suppose that the degrees of degeneracy does not change along the course of the controlled parameter $\lambda(t)$. Then the adiabatic theorem again tells that transition between different energy levels occurs with negligible probability. If the initial state is one of energy eigenstates, $|\psi(0)\rangle = |n, \alpha; \lambda(0)\rangle$, then it remains a linear combination of the instantaneous energy eigenstates with the same quantum number n:

$$|\psi(t)\rangle = \sum_{\beta=1}^{k_n} c_{\beta\alpha}(t) |n, \beta; \lambda(t)\rangle. \tag{9}$$

Since the quantum number n is unchanged, we omit it in the following equations. By substituting (9) into (2), we get

$$\frac{d}{dt} c_{\beta\alpha}(t) + \sum_{\gamma=1}^{k} \left\langle \beta; \lambda(t) \left| \frac{d}{dt} \right| \gamma; \lambda(t) \right\rangle c_{\gamma\alpha}(t) = -\frac{i}{\hbar} \varepsilon(\gamma(t)) c_{\beta\alpha}(t), \tag{10}$$

whose formal solution is given as

$$c_{\beta\alpha}(t) = \exp\left(-\frac{i}{\hbar} \int_0^t \varepsilon(\lambda(s)) ds \right) \mathcal{T} \exp\left(-\int_0^t \mathcal{A}(s) ds \right)_{\beta\alpha} \tag{11}$$

with the matrix-valued function

$$\mathcal{A}_{\beta\alpha}(t) = \left\langle \beta; \lambda(t) \left| \frac{d}{dt} \right| \alpha; \lambda(t) \right\rangle = \sum_{\mu=1}^{m} \left\langle \beta; \lambda \left| \frac{\partial}{\partial \lambda^\mu} \right| \alpha; \lambda \right\rangle \frac{d\lambda^\mu}{dt}. \tag{12}$$

The symbol $\mathcal{T}$ indicates the time-ordered product. It is easily verified that $\mathcal{A}_{\beta\alpha}^* = -\mathcal{A}_{\alpha\beta}$. We introduce a $\mathfrak{u}(k)$-valued one-form[a]

$$A_{\beta\alpha}(\lambda) = \sum_{\mu=1}^{m} A_{\beta\alpha,\mu}(\lambda) \, d\lambda^\mu = \sum_{\mu=1}^{m} \left\langle \beta; \lambda \left| \frac{\partial}{\partial \lambda^\mu} \right| \alpha; \lambda \right\rangle d\lambda^\mu \tag{13}$$

in the parameter space. A is called the Wilczek-Zee connection. Then the unitary matrix appearing in (11) is rewritten as

$$\Gamma(t) = \mathcal{P} \exp\left(-\int_{\lambda(0)}^{\lambda(t)} A \right), \tag{14}$$

where $\mathcal{P}$ stands for the path-ordered product. It is defined by the set of differential equations

$$\frac{d}{dt} \Gamma_{\beta\alpha}(t) = -\sum_{\gamma=1}^{k_n} \sum_{\mu=1}^{m} A_{\beta\gamma,\mu}(\lambda(t)) \Gamma_{\gamma\alpha}(t) \frac{d\lambda^\mu}{dt}, \qquad \Gamma_{\beta\alpha}(0) = \delta_{\beta\alpha}. \tag{15}$$

---

[a]We denote the Lie algebra of the Lie group $U(k)$ by $\mathfrak{u}(k)$, which is the set of k-dimensional skew-Hermite matrices.

We assume that the control parameter $\lambda(t)$ comes back to the initial point $\lambda(T) = \lambda(0)$. However, the state $|\psi(T)\rangle$ fails to return to the initial state and is subject to a unitary rotation as

$$|\psi(T)\rangle = \exp\left(-\frac{i}{\hbar}\int_0^T \varepsilon(\lambda(s))ds\right) \sum_{\beta=1}^k |n,\beta;\lambda(0)\rangle\Gamma_{\beta\alpha}(T). \quad (16)$$

The unitary matrix

$$\Gamma[C] := \Gamma(T) = \mathcal{P}\exp\left(-\oint_C A\right) \in U(k) \quad (17)$$

is called the *holonomy matrix* associated with the loop $\lambda(t)$. It is important to note that the holonomy $\Gamma[C]$ depends only on the shape of the loop C in the parameter space and is independent of parametrization of the loop by $\lambda(t)$.

2.4. *Examples*

Here we shall demonstrate calculations of the Berry phase and the Wilczek-Zee holonomy.

2.4.1. *Berry phase*

This example was first studied in Berry's original paper.[7] Let us consider a system of spin one half which is put in a magnetic field

$$\boldsymbol{B} = \begin{pmatrix} B_x \\ B_y \\ B_z \end{pmatrix} = B\begin{pmatrix} \sin\theta\cos\phi \\ \sin\theta\sin\phi \\ \cos\theta \end{pmatrix}. \quad (18)$$

The Hamiltonian of the spin is

$$\begin{aligned} H &= B_x\sigma_x + B_y\sigma_y + B_z\sigma_z \\ &= \begin{pmatrix} B_z & B_x - iB_y \\ B_x + iB_y & -B_z \end{pmatrix} = B\begin{pmatrix} \cos\theta & e^{-i\phi}\sin\theta \\ e^{i\phi}\sin\theta & -\cos\theta \end{pmatrix}. \end{aligned} \quad (19)$$

Its eigenvalues and eigenvectors are given as

$$H|+;\boldsymbol{B}\rangle = B|+;\boldsymbol{B}\rangle, \qquad |+;\boldsymbol{B}\rangle = \begin{pmatrix} e^{-i\phi/2}\cos(\theta/2) \\ e^{i\phi/2}\sin(\theta/2) \end{pmatrix}, \quad (20)$$

$$H|-;\boldsymbol{B}\rangle = -B|-;\boldsymbol{B}\rangle, \qquad |-;\boldsymbol{B}\rangle = \begin{pmatrix} -e^{-i\phi/2}\sin(\theta/2) \\ e^{i\phi/2}\cos(\theta/2) \end{pmatrix}. \quad (21)$$

Energy eigenvalues are nondegenerate for any nonvanishing B. Hence, by an adiabatic change of the external parameter (B,θ,ϕ), we will observe the Berry phase. The connection form $A = \langle\cdot|d|\cdot\rangle$, which is defined by (13), is calculated as follows. First, calculate the exterior derivatives of the eigenvectors to get

$$d|+; \boldsymbol{B}\rangle = d \begin{pmatrix} e^{-i\phi/2}\cos(\theta/2) \\ e^{i\phi/2}\sin(\theta/2) \end{pmatrix}$$

$$= \frac{i}{2} \begin{pmatrix} -e^{-i\phi/2}\cos(\theta/2) \\ e^{i\phi/2}\sin(\theta/2) \end{pmatrix} d\phi + \frac{1}{2} \begin{pmatrix} -e^{-i\phi/2}\sin(\theta/2) \\ e^{i\phi/2}\cos(\theta/2) \end{pmatrix} d\theta, \quad (22)$$

$$d|-; \boldsymbol{B}\rangle = d \begin{pmatrix} -e^{-i\phi/2}\sin(\theta/2) \\ e^{i\phi/2}\cos(\theta/2) \end{pmatrix}$$

$$= \frac{i}{2} \begin{pmatrix} e^{-i\phi/2}\sin(\theta/2) \\ e^{i\phi/2}\cos(\theta/2) \end{pmatrix} d\phi - \frac{1}{2} \begin{pmatrix} e^{-i\phi/2}\cos(\theta/2) \\ e^{i\phi/2}\sin(\theta/2) \end{pmatrix} d\theta. \quad (23)$$

Second, calculate the inner products

$$A_+(\boldsymbol{B}) = \langle+; \boldsymbol{B}| \, d|+; \boldsymbol{B}\rangle$$

$$= (e^{i\phi/2}\cos(\theta/2), e^{-i\phi/2}\sin(\theta/2)) \, d \begin{pmatrix} e^{-i\phi/2}\cos(\theta/2) \\ e^{i\phi/2}\sin(\theta/2) \end{pmatrix}$$

$$= (e^{i\phi/2}\cos(\theta/2), e^{-i\phi/2}\sin(\theta/2))$$
$$\left\{ \frac{i}{2} \begin{pmatrix} -e^{-i\phi/2}\cos(\theta/2) \\ e^{i\phi/2}\sin(\theta/2) \end{pmatrix} d\phi + \frac{1}{2} \begin{pmatrix} -e^{-i\phi/2}\sin(\theta/2) \\ e^{i\phi/2}\cos(\theta/2) \end{pmatrix} d\theta \right\}$$

$$= \frac{i}{2}\Big(-\cos^2(\theta/2) + \sin^2(\theta/2) \Big) d\phi$$
$$+ \frac{1}{2}\Big(-\cos(\theta/2)\sin(\theta/2) + \sin(\theta/2)\cos(\theta/2) \Big) d\theta$$

$$= -\frac{i}{2}\cos\theta \, d\phi, \quad (24)$$

$$A_-(\boldsymbol{B}) = \langle-; \boldsymbol{B}| \, d|-; \boldsymbol{B}\rangle$$

$$= (-e^{i\phi/2}\sin(\theta/2), e^{-i\phi/2}\cos(\theta/2)) \, d \begin{pmatrix} -e^{-i\phi/2}\sin(\theta/2) \\ e^{i\phi/2}\cos(\theta/2) \end{pmatrix}$$

$$= \frac{i}{2}\cos\theta \, d\phi. \quad (25)$$

For a loop $C = \partial S$ in the parameter space, we will observe the Berry phase

$$e^{i\gamma} = \exp\left[-\int_C A_+ \right] = \exp\left[\frac{i}{2}\int_C \cos\theta \, d\phi \right] = \exp\left[-\frac{i}{2}\int_S \sin\theta \, d\theta \wedge d\phi \right] \quad (26)$$

if the initial state is $|+; \boldsymbol{B}\rangle$.

2.4.2. Λ-*type system*

This system has been studied as a model of the holonomic quantum computer which uses atomic energy levels[21] and also as a model of the holonomic computer which uses Josephson junctions.[22] The Hilbert space is assumed to be $\mathbb{C}^{N+1}$. The Hamiltonian is

$$H = \varepsilon|e\rangle\langle e| + \sum_{j=1}^{N}\left(\omega_j|g_j\rangle\langle e| + \omega_j^*|e\rangle\langle g_j|\right). \tag{27}$$

The state $|e\rangle$ represents an excited state and $|g_j\rangle$ $(j = 1, \cdots, N)$ represent degenerate ground states before couplings between the states are introduced. The parameter $\varepsilon \in \mathbb{R}$ is the excitation energy and $\omega_j \in \mathbb{C}$ are coupling constants. These parameters are assumed to be controllable. The Hamiltonian can be written in the matrix form

$$H = \begin{pmatrix} \varepsilon & \omega_1^* & \omega_2^* & \cdots & \omega_N^* \\ \omega_1 & 0 & 0 & \cdots & 0 \\ \omega_2 & 0 & 0 & \cdots & 0 \\ \vdots & \vdots & \vdots & \ddots & \vdots \\ \omega_N & 0 & 0 & \cdots & 0 \end{pmatrix}. \tag{28}$$

Its eigenvalues and eigenvectors are

$$He_+ = E_+e_+, \quad e_+ = c_+\begin{pmatrix} E_+ \\ \omega_1 \\ \omega_2 \\ \vdots \\ \omega_N \end{pmatrix}, \quad E_+ = \frac{\varepsilon}{2} + \sqrt{\left(\frac{\varepsilon}{2}\right)^2 + ||\omega||^2}, \tag{29}$$

$$He_- = E_+e_-, \quad e_- = c_-\begin{pmatrix} E_- \\ \omega_1 \\ \omega_2 \\ \vdots \\ \omega_N \end{pmatrix}, \quad E_- = \frac{\varepsilon}{2} - \sqrt{\left(\frac{\varepsilon}{2}\right)^2 + ||\omega||^2}, \tag{30}$$

$$H\alpha = 0, \quad \alpha = \frac{1}{||\omega^\perp||}\begin{pmatrix} 0 \\ \omega^\perp \end{pmatrix}, \tag{31}$$

where $c_\pm = (E_\pm^2 + ||\omega||^2)^{-\frac{1}{2}}$. The vector $\omega^\perp \in \mathbb{C}^N$ satisfies $\langle\omega, \omega^\perp\rangle = 0$. Thus, the eigenvalue $E = 0$ is $(N - 1)$-fold degenerate.

When $N = 3$, we have the two-fold degenerate level $E = 0$. The eigenvectors are parametrized as

$$\alpha_1 = \frac{1}{\sqrt{|\omega_1|^2 + |\omega_2|^2}} \begin{pmatrix} 0 \\ \omega_2^* \\ -\omega_1^* \\ 0 \end{pmatrix}, \tag{32}$$

$$\alpha_2 = \frac{1}{\sqrt{(|\omega_1|^2 + |\omega_3|^2)(|\omega_1|^2 + |\omega_2|^2 + |\omega_3|^2)}} \begin{pmatrix} 0 \\ \omega_1\,\omega_3^* \\ \omega_2\,\omega_3^* \\ -(|\omega_1|^2 + |\omega_2|^2) \end{pmatrix}. \tag{33}$$

The components of the connection are to be calculated according to

$$A = \begin{pmatrix} \langle \alpha_1, d\alpha_1 \rangle & \langle \alpha_1, d\alpha_2 \rangle \\ \langle \alpha_2, d\alpha_1 \rangle & \langle \alpha_2, d\alpha_2 \rangle \end{pmatrix} = \sum_{\mu=1}^{6} A_\mu\, d\lambda^\mu \tag{34}$$

with $\omega_1 = \lambda^1 + i\lambda^2$, $\omega_2 = \lambda^3 + i\lambda^4$, $\omega_3 = \lambda^5 + i\lambda^6$. The calculation to get the concrete forms of A_μ is cumbersome and we do not show it here. When we take a loop $\lambda(t)$, the associated holonomy $g(t)$, which is an $(N-1) \times (N-1)$ matrix, is a solution of the differential equation

$$\frac{dg(t)}{dt} = -\sum_\mu A_\mu(\lambda(t))\, g(t)\, \frac{d\lambda^\mu(t)}{dt}, \qquad g(0) = I. \tag{35}$$

The concrete calculations of the holonomy can be found in literature.[21,22]

3. Holonomic Quantum Computer

Zanardi and Rasetti,[12] and also Pachos with them[13] proposed to use the Wilczek-Zee holonomy (17) as a unitary gate of quantum computer. Since the holonomy depends only on the geometry of the control-parameter loop and it is insensitive to the detail of the dynamics of the system, the holonomic quantum computer is robust against noise from the environment and tolerates incompleteness of the control. For these reasons the holonomic quantum computer is regarded as one of promising candidates for realistic quantum computers. However, a useful holonomic quantum computer should have the following properties:

(i) Controllability: a holonomic quantum computer should have a large enough number of control parameters for generating arbitrary holonomies.

(ii) Accurate degeneracy: degeneracy of the energy levels should be kept during the time-evolution. In an actual system degeneracy is often resolved by a tiny asymmetric interaction or a noise. Inaccuracy of degeneracy spoils the precision of the holonomic computation. Moreover, the degree of degeneracy is equal to the dimension of the Hilbert space available for quantum computation. Hence, to realize a large qubits system it is necessary to find a system which has a highly degenerate energy level.

(iii) Adiabaticity: we need to change the control parameter as slowly as possible to avoid undesirable transitions between different energy levels of the quantum system. No level crossings should occur during the process since a level crossing violates the adiabaticity condition.

(iv) Short execution time: an actual system cannot be completely isolated from its environment, and the interaction between the system and the environment causes loss of coherence of the quantum state of the system. We need to finish the whole process in a possible shortest time to avoid the undesirable decoherence.

The requests (i) and (ii) are practically important issues. Although many designs of working holonomic computers have been proposed, it remains a hard task to find a feasible design. The requests (iii) and (iv) can be analyzed with a mathematical method. They naturally lead us to an optimization problem; we need to find the shortest loop in the control parameter space with producing the demanded holonomy. This optimization problem is the main subject of this lecture.

4. Formulation of the Problem and Its Solution

4.1. Geometrical setting

The differential geometry of connection[3,4] provides a suitable language to describe the Wilczek-Zee holonomy. A concise review on this subject also given by Fujii.[14] We outline the geometrical setting of the problem here to make this paper self-contained.

Suppose that the state of a system is described by a finite-dimensional Hilbert space $\mathbb{C}^N$ and the system has a family of Hamiltonians acting on the Hilbert space. Assume that the ground state of each Hamiltonian is k-fold degenerate ($k < N$). A mathematical setting to describe this system is the *principal fiber bundle* $(S_{N,k}(\mathbb{C}), G_{N,k}(\mathbb{C}), \pi, U(k))$, which consists of the *Stiefel manifold* $S_{N,k}(\mathbb{C})$, the *Grassmann manifold* $G_{N,k}(\mathbb{C})$, the *projection map* $\pi : S_{N,k}(\mathbb{C}) \to G_{N,k}(\mathbb{C})$, and the unitary group $U(k)$ as explained below.

The Stiefel manifold is the set of orthonormal k-frames in $\mathbb{C}^N$,

$$S_{N,k}(\mathbb{C}) := \{V \in M(N, k; \mathbb{C}) \mid V^\dagger V = I_k\}, \tag{36}$$

where $M(N, k; \mathbb{C})$ is the set of $N \times k$ complex matrices and I_k is the k-dimensional unit matrix. The unitary group $U(k)$ acts on $S_{N,k}(\mathbb{C})$ from the right

$$S_{N,k}(\mathbb{C}) \times U(k) \to S_{N,k}(\mathbb{C}), \qquad (V, h) \mapsto Vh \tag{37}$$

by means of matrix product. It should be noted that this action is free. In other words, $h = I_k$ if there exists a point $V \in S_{N,k}(\mathbb{C})$ such that $Vh = V$.

The Grassmann manifold is defined as the set of k-dimensional subspaces in $\mathbb{C}^N$, or equivalently, as the set of projection operators of rank k,

$$G_{N,k}(\mathbb{C}) := \{P \in M(N, N; \mathbb{C}) \mid P^2 = P, \ P^\dagger = P, \ \mathrm{tr} P = k\}. \tag{38}$$

The projection map $\pi : S_{N,k}(\mathbb{C}) \to G_{N,k}(\mathbb{C})$ is defined as

$$\pi : V \mapsto P := VV^\dagger. \tag{39}$$

It is easily proven that the map π is surjective. Namely, for any $P \in G_{N,k}(\mathbb{C})$, there is $V \in S_{N,k}(\mathbb{C})$ such that $\pi(V) = P$. It is easily observed that

$$\pi(Vh) = (Vh)(Vh)^\dagger = Vhh^\dagger V^\dagger = VV^\dagger = \pi(V) \tag{40}$$

for any $h \in U(k)$ and any $V \in S_{N,k}(\mathbb{C})$. It also can be verified that if $\pi(V) = \pi(V')$ for some $V, V' \in S_{N,k}(\mathbb{C})$, there exists a unique $h \in U(k)$ such that $Vh = V'$. Thus the Stiefel manifold $S_{N,k}(\mathbb{C})$ becomes a principal bundle over $G_{N,k}(\mathbb{C})$ with the *structure group* $U(k)$.

In particular, when $k = 1$, the principal bundle $(S_{N,k}(\mathbb{C}),\ G_{N,k}(\mathbb{C}),\ \pi, U(k))$ is denoted as $(S^{2N-1}, \mathbb{C}P^{N-1}, \pi, U(1))$ and is called the *Hopf bundle*. Then the Stiefel manifold becomes a $(2N-1)$-dimensional sphere

$$S_{N,1}(\mathbb{C}) = S^{2N-1} = \{z \in \mathbb{C}^N \mid z^\dagger z = 1\}. \tag{41}$$

At this time, we can introduce an equivalence relation $z \sim z'$ for $z, z' \in S^{2N-1}$ as

$$z \sim z' :\Longleftrightarrow \exists \theta \in \mathbb{R}, \ e^{i\theta} z = z' \tag{42}$$

and denote the equivalence class of z as $[z] = \{z' \in S^{2N-1} \mid z \sim z'\} = \{e^{i\theta} z \mid \theta \in \mathbb{R}\}$. Then the quotient space $S^{2N-1}/\sim$, which is the whole set of the equivalence classes, is called the *projective space* $\mathbb{C}P^{N-1}$. Each point

$[z]$ of $\mathbb{C}P^{N-1}$ uniquely determines a one-dimensional subspace in $\mathbb{C}^N$. It is easily verified that

$$z \sim z' \iff \pi(z) = \pi(z') \tag{43}$$

for the projection map $\pi(z) = zz^\dagger$. Thus the Grassmann manifold $G_{N,1}(\mathbb{C})$ can be identified with $\mathbb{C}P^{N-1}$. The structure group $U(1) = \{e^{i\theta}\}$ acts on the fiber $[z]$. When $N = 2$ and $k = 1$, the Grassmann manifold is isomorphic to $G_{2,1}(\mathbb{C}) \cong \mathbb{C}P^1 \cong S^2$. At the same time the Stiefel manifold is $S_{2,1}(\mathbb{C}) \cong S^3$. Then the projection map is $\pi : S^3 \to S^2$.

For a general values of N and k, the group $U(N)$ acts on both $S_{N,k}(\mathbb{C})$ and $G_{N,k}(\mathbb{C})$ as

$$U(N) \times S_{N,k}(\mathbb{C}) \to S_{N,k}(\mathbb{C}), \qquad (g, V) \mapsto gV, \tag{44}$$

$$U(N) \times G_{N,k}(\mathbb{C}) \to G_{N,k}(\mathbb{C}), \qquad (g, P) \mapsto gPg^\dagger \tag{45}$$

by matrix product. It is easily verified that $\pi(gV) = g\pi(V)g^\dagger$. This action is *transitive*, namely, there is $g \in U(N)$ for any $V, V' \in S_{N,k}(\mathbb{C})$ such that $V' = gV$. There is also $g \in U(N)$ for any $P, P' \in G_{N,k}(\mathbb{C})$ such that $P' = gPg^\dagger$. The *stabilizer group* of a point $V \in S_{N,k}(\mathbb{C})$ is defined as

$$G_V := \{g \in U(N) \,|\, gV = V\} \tag{46}$$

and it is isomorphic to $U(N - k)$. Similarly, it can be verified that the stabilizer group of each point in $G_{N,k}(\mathbb{C})$ is isomorphic to $U(k) \times U(N - k)$. Thus, they form the *homogeneous bundle*

$$\pi : S_{N,k}(\mathbb{C}) \cong U(N)/U(N - k) \to G_{N,k}(\mathbb{C}) \cong U(N)/(U(k) \times U(N - k)). \tag{47}$$

The *canonical connection form* on $S_{N,k}(\mathbb{C})$ is a $\mathfrak{u}(k)$-valued one-form

$$A := V^\dagger dV. \tag{48}$$

which is a generalization of the Wilczek-Zee connection (13). This is characterized as the unique connection that is invariant under the action (44). The associated curvature two-form is then defined and calculated as

$$F := dA + A \wedge A = dV^\dagger \wedge dV + V^\dagger dV \wedge V^\dagger dV$$
$$= dV^\dagger \wedge (I_N - VV^\dagger)dV, \tag{49}$$

where we used $dV^\dagger V + V^\dagger dV = 0$, which is derived from $V^\dagger V = I_k$.

These manifolds are equipped with Riemannian metrics. We define a metric

$$\|dV\|^2 := \mathrm{tr}\,(dV^\dagger dV) \tag{50}$$

for the Stiefel manifold and

$$\|dP\|^2 := \operatorname{tr}(dPdP) \qquad (51)$$

for the Grassmann manifold.

4.2. The isoholonomic problem

Here we reformulate the Wilczek-Zee holonomy in terms of the geometric terminology introduced above. The state vector $\psi(t) \in \mathbb{C}^N$ evolves according to the Schrödinger equation

$$i\hbar \frac{d}{dt}\psi(t) = H(t)\psi(t). \qquad (52)$$

The Hamiltonian admits a spectral decomposition

$$H(t) = \sum_{n=1}^{L} \varepsilon_n(t) P_n(t) \qquad (53)$$

with projection operators $P_n(t)$. Therefore, the set of distinct energy eigenvalues $(\varepsilon_1, \dots, \varepsilon_L)$ and mutually orthogonal projectors $(P_1, \dots, P_L)$ constitutes a complete set of control parameters of the system. Now we concentrate on the eigenspace associated with the lowest energy, which is assumed to be identically zero, $\varepsilon_1 \equiv 0$. We write $P_1(t)$ as $P(t)$ simply. Suppose that the degree of degeneracy $k = \operatorname{tr} P(t)$ is constant. For each t, there exists $V(t) \in S_{N,k}(\mathbb{C})$ such that $P(t) = V(t)V^\dagger(t)$. By adiabatic approximation we mean that the state vector $\psi(t)$ belongs to the subspace spanned by the frame $V(t)$, which is the eigensubspace of the instantaneous Hamiltonian. Then for the state vector $\psi(t)$ there exists a vector $\phi(t) \in \mathbb{C}^k$ such that

$$\psi(t) = V(t)\phi(t). \qquad (54)$$

This equation is equivalent to (9). Since $H(t)\psi(t) = \varepsilon_1\psi(t) = 0$, the Schrödinger equation (52) becomes

$$\frac{d\phi}{dt} + V^\dagger \frac{dV}{dt}\phi(t) = 0 \qquad (55)$$

and its formal solution is written as

$$\phi(t) = \mathcal{P}\exp\left(-\int V^\dagger dV\right)\phi(0). \qquad (56)$$

Therefore $\psi(t)$ is written as

$$\psi(t) = V(t)\mathcal{P}\exp\left(-\int V^\dagger dV\right)V^\dagger(0)\psi(0). \qquad (57)$$

In particular, when the control parameter comes back to the initial point as $P(T) = P(0)$, the holonomy $\Gamma \in U(k)$ is defined via

$$\psi(T) = V(0)\Gamma\,\phi(0) \tag{58}$$

and is given explicitly as

$$\Gamma = V(0)^\dagger\,V(T)\,\mathcal{P}\exp\left(-\int V^\dagger dV\right). \tag{59}$$

If the condition

$$V^\dagger \frac{dV}{dt} = 0 \tag{60}$$

is satisfied, the curve $V(t)$ in $S_{N,k}(\mathbb{C})$ is called a *horizontal lift* of the curve $P(t) = \pi(V(t))$ in $G_{N,k}(\mathbb{C})$. Then the holonomy (59) is reduced to

$$\Gamma = V^\dagger(0)V(T) \in U(k). \tag{61}$$

Now we are ready to state the *isoholonomic problem* in the present context; given a specified unitary gate $U_{\text{gate}} \in U(k)$ and an initial point $P_0 \in G_{N,k}(\mathbb{C})$, find the shortest loop $P(t)$ in $G_{N,k}(\mathbb{C})$ with the base points $P(0) = P(T) = P_0$ whose horizontal lift $V(t)$ in $S_{N,k}(\mathbb{C})$ produces a holonomy Γ that coincides with U_{gate}. This problem was first investigated systematically by a mathematician, Montgomery,[15] who was motivated from experimental studies of Berry phase.

The isoholonomic problem is formulated as a variational problem. The length of the horizontal curve $V(t)$ is evaluated by the functional

$$S[V,\Omega] = \int_0^T \left\{ \operatorname{tr}\left(\frac{dV^\dagger}{dt}\frac{dV}{dt}\right) - \operatorname{tr}\left(\Omega\,V^\dagger\frac{dV}{dt}\right) \right\} dt, \tag{62}$$

where $\Omega(t) \in \mathfrak{u}(k)$ is a Lagrange multiplier to impose the horizontality condition (60) on the curve $V(t)$. Note that the value of the functional S is equal to the length of the projected curve $P(t) = \pi(V(t))$,

$$S = \int_0^T \frac{1}{2}\operatorname{tr}\left(\frac{dP}{dt}\frac{dP}{dt}\right) dt. \tag{63}$$

Thus the problem is formulated as follows; find functions $V(t)$ and $\Omega(t)$ that attain an extremal value of the functional (62) and satisfy the boundary condition (61).

4.3. The solution: horizontal extremal curve

We shall derive the Euler-Lagrange equation associated to the functional (62) and solve it explicitly. A variation of the curve $V(t)$ is defined by an arbitrary smooth function $\eta(t) \in \mathfrak{u}(N)$ such that $\eta(0) = \eta(T) = 0$ and an infinitesimal parameter $\epsilon \in \mathbb{R}$ as

$$V_\epsilon(t) = (1 + \epsilon\eta(t))V(t). \tag{64}$$

By substituting $V_\epsilon(t)$ into (62) and differentiating with respect to ϵ, the extremal condition yields

$$\begin{aligned}
0 = \left. \frac{dS}{d\epsilon}\right|_{\epsilon=0} &= \int_0^T \operatorname{tr}\left\{\dot{\eta}(V\dot{V}^\dagger - \dot{V}V^\dagger - V\Omega V^\dagger)\right\} dt \\
&= \left[\operatorname{tr}\left\{\eta(V\dot{V}^\dagger - \dot{V}V^\dagger - V\Omega V^\dagger)\right\}\right]_{t=0}^{t=T} \\
&\quad - \int_0^1 \operatorname{tr}\left\{\eta\frac{d}{dt}(V\dot{V}^\dagger - \dot{V}V^\dagger - V\Omega V^\dagger)\right\} dt.
\end{aligned} \tag{65}$$

Thus we obtain the Euler-Lagrange equation

$$\frac{d}{dt}(\dot{V}V^\dagger - V\dot{V}^\dagger + V\Omega V^\dagger) = 0. \tag{66}$$

The horizontality condition $V^\dagger\dot{V} = 0$ is re-derived from the extremal condition with respect to $\Omega(t)$. Finally, the isoholonomic problem is reduced to the set of equations (60) and (66), which we call a *horizontal extremal equation*.

Next, we solve the equations (60) and (66). The equation (66) is integrable and yields

$$\dot{V}V^\dagger - V\dot{V}^\dagger + V\Omega V^\dagger = \text{const} = X \in \mathfrak{u}(N). \tag{67}$$

Conjugation of the horizontality condition (60) yields $\dot{V}^\dagger V = 0$. Then, by multiplying V on (67) from the right we obtain

$$\dot{V} + V\Omega = XV. \tag{68}$$

By multiplying $V^\dagger$ on (68) from the left we obtain

$$\Omega = V^\dagger XV. \tag{69}$$

The equation (68) implies $\dot{V} = XV - V\Omega$, and hence the time derivative of

$\Omega(t)$ becomes

$$\begin{aligned}
\dot{\Omega} &= V^{\dagger} X \dot{V} + \dot{V}^{\dagger} X V \\
&= V^{\dagger} X (XV - V\Omega) + (-V^{\dagger} X + \Omega V^{\dagger}) X V \\
&= V^{\dagger} X X V - \Omega \Omega - V^{\dagger} X X V + \Omega \Omega \\
&= 0.
\end{aligned} \tag{70}$$

We used the facts that $X^{\dagger} = -X$ and $\Omega^{\dagger} = -\Omega$. Therefore, $\Omega(t)$ is actually a constant. Thus the solution of (68) and (69) is

$$V(t) = e^{tX} V_0 \, e^{-t\Omega}, \qquad \Omega = V_0^{\dagger} X V_0. \tag{71}$$

We call this solution the *horizontal extremal curve*. Then (67) becomes

$$(XV - V\Omega) V^{\dagger} - V(-V^{\dagger} X + \Omega V^{\dagger}) + V\Omega V^{\dagger} = X,$$

which is arranged as

$$X - (VV^{\dagger} X + XVV^{\dagger} - VV^{\dagger} XVV^{\dagger}) = 0, \tag{72}$$

where we used (69). Since the group $U(N)$ acts on $S_{N,k}(\mathbb{C})$ freely, we may take

$$V_0 = \begin{pmatrix} I_k \\ 0 \end{pmatrix} \in S_{N,k}(\mathbb{C}) \tag{73}$$

as the initial point without loss of generality. We can parametrize $X \in \mathfrak{u}(N)$, which satisfies (69), as

$$X = \begin{pmatrix} \Omega & W \\ -W^{\dagger} & Z \end{pmatrix} \tag{74}$$

with $W \in M(k, N - k; \mathbb{C})$ and $Z \in \mathfrak{u}(N - k)$. Then the constraint equation (72) forces us to choose

$$Z = 0. \tag{75}$$

Via the above discussion we obtained a complete set of solution (71) of the horizontal extremal equation (60) and (66). The complete solutions are parametrized with constant matrices $\Omega \in \mathfrak{u}(k)$ and $W \in M(k, N - k; \mathbb{C})$, and are written explicitly as

$$V(t) = e^{tX} V_0 \, e^{-t\Omega}, \qquad X = \begin{pmatrix} \Omega & W \\ -W^{\dagger} & 0 \end{pmatrix}. \tag{76}$$

We call the matrix X a *controller*. At this time the holonomy (61) is expressed as

$$\Gamma = V^{\dagger}(0) V(T) = V_0^{\dagger} \, e^{TX} V_0 \, e^{-T\Omega} \in U(k). \tag{77}$$

These results (76) and (77) coincide with Montgomery's results[b]. We took a different approach from his; in the above discussion we wrote down the Euler-Lagrange equation and solved it directly.

We evaluate the length of the extremal curve for later convenience by substituting (76) into (63) as

$$S = \int_0^T \frac{1}{2} \mathrm{tr} \left(\frac{dP}{dt} \frac{dP}{dt} \right) dt = \mathrm{tr} \left(W^\dagger W \right) T. \tag{78}$$

5. The Boundary-Value Problem

Remember that we are seeking for the optimally shortest loop in the Grassmann manifold which generates a desired holonomy. The optimality is guaranteed by solving the horizontal extremal equation. The remaining task is to find a solution that satisfies two boundary conditions. First, when the curve $V(t)$ in $S_{N,k}(\mathbb{C})$ is projected into $G_{N,k}(\mathbb{C})$ via the map $\pi : S_{N,k}(\mathbb{C}) \to G_{N,k}(\mathbb{C}), V \mapsto VV^\dagger$, it should make a loop. Thus the curve should satisfy the *loop condition*

$$V(T)V^\dagger(T) = e^{TX} V_0 V_0^\dagger e^{-TX} = V_0 V_0^\dagger. \tag{79}$$

Second, the terminal points $V(0)$ and $V(T)$ of the curve $V(t)$ define the holonomy $\Gamma = V^\dagger(0)V(T)$ as in (61). Thus the curve should satisfy the *holonomy condition*

$$V_0^\dagger V(T) = V_0^\dagger e^{TX} V_0 e^{-T\Omega} = U_{\mathrm{gate}} \tag{80}$$

for a specified unitary gate $U_{\mathrm{gate}} \in U(k)$.

We need to find the suitable parameters Ω and W in solutions (76) satisfying the above two conditions. Montgomery[15] presented this boundary-value problem as an open problem. In this section we give a scheme to solve systematically this problem.

5.1. *Equivalence class*

There is a class of equivalent solutions for a given initial condition V_0 and a given final condition $V(T) = V_0 U_{\mathrm{gate}}$. Here we clarify the equivalence relation among solutions $\{V(t)\}$ that have the form (76) and satisfy (79) and (80).

---

[b]In his paper[15] Montgomery cited Bär's theorem to complete the proof. However, Bär's paper being a diploma thesis, it is not widely available. Instead of relying on Bär's theorem, here we did the explicit calculation to reach the same result.

We say that two solutions $V(t)$ and $V'(t)$ are equivalent if there are elements $g \in U(N)$ and $h \in U(k)$ such that $V(t)$ and

$$V'(t) = gV(t)h^\dagger \tag{81}$$

satisfy the same boundary conditions

$$gV_0 h^\dagger = V_0 \tag{82}$$

and

$$hU_{\text{gate}}h^\dagger = U_{\text{gate}}. \tag{83}$$

For the initial point (73), the condition (82) states that $g \in U(N)$ must have a block-diagonal form

$$g = \begin{pmatrix} h_1 & 0 \\ 0 & h_2 \end{pmatrix}, \qquad h = h_1 \in U(k), \quad h_2 \in U(N-k). \tag{84}$$

The controller X' of $V'(t)$ are then found from

$$V'(t) = gV(t)h^\dagger = ge^{tX}g^\dagger gV_0 h^\dagger h e^{-t\Omega}h^\dagger = e^{tgXg^\dagger}gV_0 h^\dagger e^{-th\Omega h^\dagger}$$
$$= e^{tgXg^\dagger}V_0 \, e^{-th\Omega h^\dagger}. \tag{85}$$

In summary, two controllers X and X' are equivalent if and only if there are unitary matrices $h_1 \in U(k)$ and $h_2 \in U(N-k)$ such that

$$X = \begin{pmatrix} \Omega & W \\ -W^\dagger & 0 \end{pmatrix}, \quad X' = \begin{pmatrix} h_1 \Omega h_1^\dagger & h_1 W h_2^\dagger \\ -h_2 W^\dagger h_1^\dagger & 0 \end{pmatrix}, \quad h_1 U_{\text{gate}} h_1^\dagger = U_{\text{gate}}. \tag{86}$$

5.2. $U(1)$ holonomy

Here we calculate the holonomy for the case $N = 2$ and $k = 1$. In this case the homogeneous bundle $\pi : S_{2,1}(\mathbb{C}) \to G_{2,1}(\mathbb{C})$ is the Hopf bundle $\pi : S^3 \to S^2$ with the structure group $U(1)$ and the Wilczek-Zee holonomy reduces to the Berry phase. In the subsequent subsection we will use the Berry phase as a building block to construct a non-Abelian holonomy. We normalize the cycle time as $T = 1$ in the following. Using real numbers $w_1, w_2, w_3 \in \mathbb{R}$ we parametrize the controller as

$$X = \begin{pmatrix} 2iw_3 & iw_1 + w_2 \\ iw_1 - w_2 & 0 \end{pmatrix} = iw_3 I + iw_1\sigma_1 + iw_2\sigma_2 + iw_3\sigma_3, \tag{87}$$

where $\boldsymbol{\sigma} = (\sigma_1, \sigma_2, \sigma_3) = (\sigma_x, \sigma_y, \sigma_z)$ are the Pauli matrices which have been introduced at (19). The exponentiation of X yields

$$e^{tX} = e^{itw_3}(I\cos\rho t + i\boldsymbol{n} \cdot \boldsymbol{\sigma}\sin\rho t), \tag{88}$$

where ρ and $\boldsymbol{n}$ are defined as

$$\rho := \|\boldsymbol{w}\| = \sqrt{(w_1)^2 + (w_2)^2 + (w_3)^2}, \qquad \boldsymbol{w} = \|\boldsymbol{w}\|\boldsymbol{n}. \qquad (89)$$

The associated horizontal extremal curve (71) then becomes

$$V(t) = e^{tX} V_0 e^{-t\Omega} = e^{-itw_3} \begin{pmatrix} \cos\rho t + in_3 \sin\rho t \\ (in_1 - n_2)\sin\rho t \end{pmatrix} \qquad (90)$$

and the projected curve in S^2 becomes

$$\begin{aligned}
P(t) &= V(t)V^\dagger(t) \\
&= \frac{1}{2}I + \frac{1}{2}\boldsymbol{\sigma} \cdot \Big[\boldsymbol{n}(\boldsymbol{n}\cdot\boldsymbol{e}_3) + (\boldsymbol{e}_3 - \boldsymbol{n}(\boldsymbol{n}\cdot\boldsymbol{e}_3))\cos 2\rho t \\
&\qquad\qquad - (\boldsymbol{n}\times\boldsymbol{e}_3)\sin 2\rho t \Big],
\end{aligned} \qquad (91)$$

where $\boldsymbol{e}_3 = (0,0,1)$. We see from (91) that the point $P(t)$ in S^2 starts at the north pole $\boldsymbol{e}_3$ of the sphere and moves along a *small circle* with the axis $\boldsymbol{n}$ in the clockwise sense by the angle $2\rho t$. The point $P(t)$ comes back to the north pole when t satisfies $2\rho t = 2\pi n$ with an integer n. To make a loop, namely, to satisfy the loop condition (79) at $t = T = 1$, the control parameters must satisfy

$$\rho = \|\boldsymbol{w}\| = n\pi \qquad (n = \pm 1, \pm 2, \dots). \qquad (92)$$

Then, the point $P(t)$ travels the same small circle n times during $0 \le t \le 1$. Therefore, the integer n counts the winding number of the loop. At $t = 1$, $\cos\rho = (-1)^n$ and the holonomy (80) is evaluated as

$$V_0^\dagger e^X V_0 e^{-\Omega} = e^{iw_3}(-1)^n e^{-2iw_3} = e^{-i(w_3 - n\pi)} = U_{\text{gate}} = e^{i\gamma}. \qquad (93)$$

Thus, to produce the holonomy $U_{\text{gate}} = e^{i\gamma}$, the controller parameters are fixed as

$$w_3 = n\pi - \gamma, \qquad w_1 + iw_2 = e^{-i\phi}\sqrt{(n\pi)^2 - (n\pi - \gamma)^2}. \qquad (94)$$

This is the solution to the boundary problem defined by (79) and (80). Here the nonvanishing integer n must satisfy $(n\pi)^2 - (n\pi - \gamma)^2 > 0$. The real parameter ϕ is not fixed by the loop condition and the holonomy condition. The phase $h_2 = e^{i\phi}$ parametrizes solutions in an equivalence class as observed in (86). Hence each equivalence class is characterized by the integer n.

The length of the loop, (78), is now evaluated as

$$S = \text{tr}\,(W^\dagger W)\,T = (n\pi)^2 - (n\pi - \gamma)^2 = \gamma(2n\pi - \gamma). \qquad (95)$$

For a fixed γ in the range $0 \leq \gamma < 2\pi$, the simple loop with $n = 1$ is the shortest one among the extremal loops. Thus, we conclude that the controller of $U_{\text{gate}} = e^{i\gamma}$ is

$$X = \begin{pmatrix} \dfrac{2i(\pi - \gamma)}{} & ie^{i\phi}\sqrt{\pi^2 - (\pi - \gamma)^2} \\ ie^{-i\phi}\sqrt{\pi^2 - (\pi - \gamma)^2} & 0 \end{pmatrix}. \tag{96}$$

We call this solution a *small circle solution* because of its geometric picture mentioned above.

5.3. $U(k)$ holonomy

Here we give a prescription to construct a controller matrix X that generates a specific unitary gate U_{gate}. It turns out that the working space should have a dimension $N \geq 2k$ to apply our method. In the following we assume that $N = 2k$. The time interval is normalized as $T = 1$ as before.

Our method consists of three steps: first, diagonalize the unitary matrix U_{gate} to be implemented, second, construct a diagonal controller matrix by combining small circle solutions, third, undo diagonalization of the controller.

In the first step, we diagonalize a given unitary matrix $U_{\text{gate}} \in U(k)$ as

$$R^\dagger U_{\text{gate}} R = U_{\text{diag}} = \text{diag}(e^{i\gamma_1}, \ldots, e^{i\gamma_k}) \tag{97}$$

with $R \in U(k)$. Each eigenvalue γ_j is taken in the range $0 \leq \gamma_j < 2\pi$. In the second step, we combine single loop solutions associated with the Berry phase to construct two $k \times k$ matrices

$$\Omega_{\text{diag}} = \text{diag}(i\omega_1, \ldots, i\omega_k), \qquad \omega_j = 2(\pi - \gamma_j), \tag{98}$$

$$W_{\text{diag}} = \text{diag}(i\tau_1, \ldots, i\tau_k), \qquad \tau_j = e^{i\phi_j}\sqrt{\pi^2 - (\pi - \gamma_j)^2}. \tag{99}$$

Then we obtain a diagonal controller

$$X_{\text{diag}} = \begin{pmatrix} \Omega_{\text{diag}} & W_{\text{diag}} \\ -W_{\text{diag}}^\dagger & 0 \end{pmatrix}.$$

In the third step, we construct the controller X as

$$X = \begin{pmatrix} R & 0 \\ 0 & I_k \end{pmatrix} \begin{pmatrix} \Omega_{\text{diag}} & W_{\text{diag}} \\ -W_{\text{diag}}^\dagger & 0 \end{pmatrix} \begin{pmatrix} R^\dagger & 0 \\ 0 & I_k \end{pmatrix} = \begin{pmatrix} R\Omega_{\text{diag}}R^\dagger & RW_{\text{diag}} \\ -W_{\text{diag}}^\dagger R^\dagger & 0 \end{pmatrix}, \tag{100}$$

which is a $2k \times 2k$ matrix. We call the set of equations, (97), (98), (99) and (100), *constructing equations of the controller*. This is our main result.

The diagonal controller X_{diag} is actually a direct sum of controllers (96), which generate Berry phases $\{e^{i\gamma_j}\}$. Hence, its holonomy is also a direct sum of the Berry phases (93) as

$$V_0^\dagger \, e^{X_{\text{diag}}} \, V_0 \, e^{-\Omega_{\text{diag}}} = U_{\text{diag}}$$

and hence we have

$$V_0^\dagger \, e^{X} \, V_0 \, e^{-\Omega} = R V_0^\dagger \, e^{X_{\text{diag}}} \, V_0 \, R^\dagger R \, e^{-\Omega_{\text{diag}}} R^\dagger = R U_{\text{diag}} R^\dagger = U_{\text{gate}}.$$

6. Examples of Unitary Gates

To illustrate our method here we construct controllers for the one-qubit Hadamard gate and the two-qubit discrete Fourier transformation (DFT) gate. For each unitary gate U_{gate}, we need to calculate the diagonalizing matrix R. Then the constructing equations of the controller, (97)-(100), provide the desired optimal controller matrices.

6.1. *Hadamard gate*

The Hadamard gate is a one-qubit gate defined as

$$U_{\text{Had}} = \frac{1}{\sqrt{2}} \begin{pmatrix} 1 & 1 \\ 1 & -1 \end{pmatrix}. \tag{101}$$

It is diagonalized by

$$R = \begin{pmatrix} \cos\frac{\pi}{8} & -\sin\frac{\pi}{8} \\ \sin\frac{\pi}{8} & \cos\frac{\pi}{8} \end{pmatrix}, \quad \cos\frac{\pi}{8} = \frac{\sqrt{2+\sqrt{2}}}{2}, \quad \sin\frac{\pi}{8} = \frac{\sqrt{2-\sqrt{2}}}{2} \tag{102}$$

as

$$R^\dagger U_{\text{Had}} R = \begin{pmatrix} 1 & 0 \\ 0 & -1 \end{pmatrix}. \tag{103}$$

Therefore, we have $\gamma_1 = 0$ and $\gamma_2 = \pi$. We may put $\phi_1 = \phi_2 = 0$. The ingredients of the constructing equations of the controller, (97)-(100), are

$$\Omega_{\text{diag}} = \text{diag}(2i\pi, 0), \quad W_{\text{diag}} = \text{diag}(0, i\pi), \tag{104}$$

and hence

$$R\Omega_{\text{diag}}R^\dagger = \frac{i\pi}{\sqrt{2}} \begin{pmatrix} \sqrt{2}+1 & 1 \\ 1 & \sqrt{2}-1 \end{pmatrix}, \quad RW_{\text{diag}} = \frac{i\pi}{2} \begin{pmatrix} 0 & -\sqrt{2-\sqrt{2}} \\ 0 & \sqrt{2+\sqrt{2}} \end{pmatrix}. \tag{105}$$

Substituting these into (100), we obtain the optimal controller of the Hadamard gate.

6.2. DFT2 gate

Discrete Fourier transformation (DFT) gates are important in many quantum algorithms including Shor's algorithm for integer factorization. The two-qubit DFT (DFT2) is a unitary transformation

$$U_{\text{DFT2}} = \frac{1}{2} \begin{pmatrix} 1 & 1 & 1 & 1 \\ 1 & i & -1 & -i \\ 1 & -1 & 1 & -1 \\ 1 & -i & -1 & i \end{pmatrix}. \tag{106}$$

It is diagonalized by

$$R = \frac{1}{2} \begin{pmatrix} 1 & \sqrt{2} & -1 & 0 \\ 1 & 0 & 1 & -\sqrt{2} \\ -1 & \sqrt{2} & 1 & 0 \\ 1 & 0 & 1 & \sqrt{2} \end{pmatrix} \tag{107}$$

as

$$R^\dagger U_{\text{DFT2}} R = \begin{pmatrix} 1 & 0 & 0 & 0 \\ 0 & 1 & 0 & 0 \\ 0 & 0 & -1 & 0 \\ 0 & 0 & 0 & i \end{pmatrix}. \tag{108}$$

Therefore, we have $\gamma_1 = \gamma_2 = 0$, $\gamma_3 = \pi$ and $\gamma_4 = \pi/2$. Thus the ingredients of the controller are

$$\Omega_{\text{diag}} = \text{diag}(2i\pi, 2i\pi, 0, i\pi), \qquad W_{\text{diag}} = \text{diag}(0, 0, i\pi, i\pi\sqrt{3}/2), \tag{109}$$

and hence

$$R\Omega_{\text{diag}}R^\dagger = \frac{i\pi}{2} \begin{pmatrix} 3 & 1 & 1 & 1 \\ 1 & 2 & -1 & 0 \\ 1 & -1 & 3 & -1 \\ 1 & 0 & -1 & 2 \end{pmatrix}, \qquad RW_{\text{diag}} = \frac{i\pi}{2} \begin{pmatrix} 0 & 0 & -1 & 0 \\ 0 & 0 & 1 & -\sqrt{3/2} \\ 0 & 0 & 1 & 0 \\ 0 & 0 & 1 & \sqrt{3/2} \end{pmatrix}. \tag{110}$$

Substituting these into (100), we finally obtain the optimal controller of the DFT2 gate.

7. Discussions

Although we provided a concrete prescription to calculate a controller of the holonomic quantum computer to execute an arbitrary unitary gate, there remain several problems for building a practical holonomic quantum computer. In the rest of this lecture we shall discuss these issues.

7.1. *Restricted control parameters*

To solve the isoholonomic problem we assumed that the Stiefel manifold is a control manifold. In other words, we assumed that we can change the projection oprator $P_1(t)$ appearing in the Hamiltonian (53) arbitrarily. However, realistic systems do not allow such a large controllability. Usually a real system admits only a smaller control manifold; the number of control parameters can be less and the ranges of their allowed values can be narrow. Finding and optimizing a loop in the control manifold to generate a demanded holonomy become a difficult task for a realistic system.

7.2. *Implementation*

Here we describe a (not complete) list of proposals for implementing holonomic quantum computers although most of them still remain theoretical proposals. Mead[16] proposed a use of Kramers degeneracy in molecules to observe the non-Abelian holonomy. Zee[17] proposed a use of nuclear quadrupole resonance. Unanyan et al.[18] proposed a use of four-level atoms manipulated with laser as a holonomic quantum computer. Jones et al.[19] implemented geometric quantum computation with NMR. Duan et al.[20] proposed a use of trapped ions. Recati et al.[21] proposed a use of neutral atoms and the photon field in cavity. Faoro et al.[22] proposed a use of Josephson junction circuits. Solinas et al.[23] studied the holonomy in a semiconductor quantum dots.

On the other hand, a lot of theoretical studies related to holonomic quantum computation can be found in literature. Karle and Pachos[24] discussed holonomy in the Grassmann manifold, which is a special case of our study. Niskanen et al.[25,26] studied numerical optimization of holonomic quantum computer. Zhu and Zanardi[27] studied robustness holonomic quantum computation against control errors. Robustness against decoherence has been analyzed, for example, by Fuentes-Guridi et al.[28] and a use of decoherence-free subspace also has been studied by Wu et al.[29]

However, the most important issue for making a practical holonomic quantum computer is to build a controllable system that has a degenerate energy level; the degeneracy must be precise and the degenerate subspace must have large dimensions to be used as a multi-qubit processor. It seems necessary to find a break-through in this direction.

Acknowledgements

The exact solution of the isoholonomic problem, which was explained in this lecture, is a result of collaborations with Daisuke Hayashi and Mikio Nakahara. I would like to thank Mikio Nakahara for stimulating discussions with him. I also appreciate efforts of the organizers of this successful summer school, Mathematical Aspects of Quantum Computing, held at Kinki University in August, 2007. This work is financially supported by Japan Society for the Promotion of Science, Grant Nos. 15540277 and 17540372.

References

1. S. Tanimura, D. Hayashi, and M. Nakahara, *Phys. Lett. A* **325** (2004) 199.
2. S. Tanimura, M. Nakahara, and D. Hayashi, *J. Math. Phys.* **46** (2005) 022101.
3. S. Kobayashi and K. Nomizu, *Foundations of Differential Geometry*, Vol. 2 (Interscience Publishers, New York, 1969).
4. M. Nakahara, *Geometry, Topology and Physics*, 2nd ed. (IOP Publishing, Bristol and New York, 2003).
5. T. Kato, *J. Phys. Soc. Japan* **5** (1950) 435.
6. A. Messiah, *Quantum Mechanics* (Dover, New York, 2000).
7. M. V. Berry, *Proc. R. Soc. Lond. A* **392** (1984) 45.
8. A. Shapere and F. Wilczek (editors), *Geometric Phases in Physics* (World Scientific, Singapore, 1989).
9. S. Tanimura, *Tom and Berry* (A thesis paper on holonomies in dynamical systems written in Japanese), *Soryushiron Kenkyu* **85** (1992) 1.
10. A. Bohm, A. Mostafazadeh, H. Koizumi, Q. Niu, and J. Zwanziger, *The Geometric Phase in Quantum Systems* (Springer, 2003).
11. F. Wilczek and A. Zee, *Phys. Rev. Lett.* **52** (1984) 2111.
12. P. Zanardi and M. Rasetti, *Phys. Lett. A* **264** (1999) 94.
13. J. Pachos, P. Zanardi, and M. Rasetti, *Phys. Rev. A* **61** (1999) 010305(R).
14. K. Fujii, *J. Math. Phys.* **41** (2000) 4406.
15. R. Montgomery, *Commun. Math. Phys.* **128** (1991) 565.
16. C. A. Mead, *Phys. Rev. Lett.* **59** (1987) 161.
17. A. Zee, *Phys. Rev. A* **38** (1988) 1.
18. R. G. Unanyan, B. W. Shore, and K. Bergmann, *Phys. Rev. A* **59** (1999) 2910.
19. J. Jones, V. Vedral, A. K. Ekert, and C. Castagnoli, *Nature* **403** (2000) 869.
20. L.-M. Duan, J. I. Cirac, and P. Soller, *Science* **292** (2001) 1695.
21. A. Recati, T. Calarco, P. Zanardi, J. I. Cirac, and P. Zoller, *Phys. Rev. A* **66** (2002) 032309.
22. L. Faoro, J. Siewert, and R. Fazio, *Phys. Rev. Lett.* **90** (2003) 28301.
23. P. Solinas, P. Zanardi, N. Zanghi, and F. Rossi, *Phys. Rev. B* **67** (2003) 121307.
24. R. Karle and J. Pachos, *J. Math. Phys.* **44** (2003) 2463.
25. A. O. Niskanen, M. Nakahara, and M. M. Salomaa, *Quantum Inf. Comput.* **2** (2002) 560.

26. A. O. Niskanen, M. Nakahara, and M. M. Salomaa, *Phys. Rev. A* **67** (2003) 012319.
27. S.-L. Zhu and P. Zanardi, *Phys. Rev. A* **72** (2005) 020301.
28. I. Fuentes-Guridi, F. Girelli, and E. Livine, *Phys. Rev. Lett.* **94** (2005) 020503.
29. L.-A. Wu, P. Zanardi, and D. A. Lidar, *Phys. Rev. Lett.* **95** (2005) 130501.

PLAYING GAMES IN QUANTUM MECHANICAL SETTINGS: FEATURES OF QUANTUM GAMES

ŞAHIN KAYA ÖZDEMIR, JUNICHI SHIMAMURA, NOBUYUKI IMOTO

Division of Materials Physics, Department of Materials Engineering Science, Graduate School of Engineering Science, Osaka University, Toyonaka, Osaka 560-8531, Japan

SORST Research Team for Interacting Carrier Electronics, 4-1-8 Honmachi, Kawaguchi, Saitama 331-0012, Japan

CREST Photonic Quantum Information Project, 4-1-8 Honmachi, Kawaguchi, Saitama 331-0012, Japan

E-mail: ozdemir@qi.mp.es.osaka-u.ac.jp

In this lecture note, we present the implications of playing classical games in quantum mechanical settings where the quantum mechanical toolbox consisting of entanglement, quantum operations and measurement is used. After a brief introduction to the concepts of classical game theory and quantum mechanics, we study quantum games and their corresponding classical analogues to determine the novelties. In addition, we introduce a benchmark which attempts to make a fair comparison of classical games and their quantum extensions. This benchmark exploits the fact that in special settings a classical game should be reproduced as a subgame of its quantum extension. We obtained a rather surprising result that this requirement prevents the use of a large set of entangled states in quantum extension of classical games.

Keywords: Game Theory, Quantum Mechanics, Entanglement, Quantum Games.

1. Introduction

Game theory is a mathematical modelling tool to formally describe conflict, cooperation or competition situations among rational and intelligent decision makers who have different preferences. Each of the decision makers is motivated to increase his/her utility which is not solely determined by his/her action but by the interaction of the strategies of all the decision makers.[1]

The start of modern game theory dates back to 1928 when J. von Neumann defined and completely solved two-person zero-sum games.[2] Later on, von Neumann introduced the concepts of linear programming and the fundamentals of game theory which later became the building blocks for his book with Morgenstern.[3,4] Although von Neumann made important contributions to quantum mechanics and game theory around the same years, the two were developed as separate fields with no or very little interaction, and with no common ground for applications. His attempts to elucidate the economical problems using physical examples faced strong criticism based on the argument that social sciences cannot be modelled after physics because the social sciences should take psychology into account. J. von Neumann's response to these criticism was "such statements are at least premature.".[3] Critiques, which say physics and social sciences are distinct fields and should not mix, are still alive, and so is the response of von Neumann.

Game theory and its theoretical toolbox have been traditionally limited to economical and evolutionary biology problems. However, recently mathematical models and techniques of game theory have increasingly been used by computer and information scientists,[5-11] Game theoretical models of economics, evolutionary biology, population dynamics, large scale distribution systems, learning models and resource allocation problems have been studied. Proposals have been done for modelling complex physical phenomena such as decoherence and irreversibility using game theoretical models.[12,13] If we bear in mind that a game is about the transfer of information, then it is not a surprise that all those distinct areas meet in game theory: Information constitutes a common ground and a strong connection among them.[14] Then one naturally wonders what happens if the information carriers are taken to be quantum systems. This leads to the introduction of the quantum mechanical toolbox into game theory to combine two distinct interests of von Neumann in an environment to mutually benefit from both, and to see what new features will arise from this combination.

Quantum mechanical toolbox (unitary operators, superposition and entangled states, measurement) was introduced in game theory, for the first time, by Meyer[15] who showed that in a coin tossing game a player utilizing quantum superposition could win with certainty against a classical player. Following this first paper, Eisert et al. studied the well-known Prisoners' dilemma (PD) using a general protocol for two player-two strategy games where the players share entanglement.[16] These two works raised an excitement in the quantum mechanics community mostly due to emergence of a new field where quantum mechanics shows its advantage and also due to the hope of using game theoretical toolbox to study quantum information tasks and quantum algorithms. Since then most of the works has been devoted to the analysis of dilemma containing classical game dynamics using Eisert's scheme and its multi-party extension. Although some quantum communication protocols and algorithms[17,18] were reformulated in the language of game theory, this area still begs for more exploration. In the mean time, quantum versions of some important theories, e.g., min-max theorem, have been developed.[19]

In this review, we will briefly introduce the terminology and basic concepts of classical game theory. Then we will discuss Eisert's scheme and focus on game dynamics with respect to the changes in the quantum mechanical resources used. We will derive a condition that should be satisfied for an entangled state to be useful in quantum version of classical games. Finally, we will outline the results.

2. A Brief Review of Classical Game Theory

The ingredients of a game theory problem include:[1] **(i)** *Players:* Two or more goal oriented, rational and intelligent decision makers. **(ii)** *Strategies:* Feasible action space of players. Players can choose to play either in pure or mixed strategies. In pure strategy, each player chooses an action with unit probability, whereas in mixed strategy the players randomize among their actions. **(iii)** *Payoffs:* Benefits, prizes, or awards distributed to players depending on the choices of all players. Payoff to a player is determined by the joint action of all players.

Rationality of players and *common knowledge* are two important assumptions in game theory. Rationality implies that each player aims at maximizing his own payoff, and that all rational players search for an equilibrium on which all can settle down. The common knowledge assumption requires that all the players know the rules of the game and that all the players know that all players know the rules of the game.

In game theory, games are generally modelled either in *strategic form* or in *extensive form*. If the players take their actions simultaneously or without observing the actions taken by other players, the *strategic form*, which is best represented by a payoff matrix, is preferred. On the other hand, if the players take their actions sequentially after observing the actions taken by the players before them, it is best to model using *extensive form*, which is represented by a game tree. The extensive form requires more specifications such as *which player* moves *when* doing *what* and with *what information*. In this work, we consider only the games with strategic form.

Games are usually grouped as *cooperative* or *noncooperative* games. In the first one, players are allowed to form coalitions and combine their decision making problems with binding commitments, whereas in the latter

Table 1. Two-player two-strategy game in strategic form.

	Bob: B_1	Bob: B_2
Alice: A_1	(a, w)	(b, x)
Alice: A_2	(c, y)	(d, z)

each player pursues his/her own interests with no intention to negotiate.

Games can also be classified according to the structure of their payoff matrices as *symmetric* or *asymmetric, zero sum* or *non-zero sum,* and *coordination* (at least one equilibrium in pure strategies) or *discoordination* games. In Table 1, payoff matrix of a generic two-player two-strategy game is given. The players, Alice (row player) and Bob (column player), have the action sets $\{A_1, A_2\}$ and $\{B_1, B_2\}$. The payoffs are represented as $(.., ..)$ with the first element being Alice's payoff and the second Bob's. This payoff matrix represents a symmetric game if $a = w, b = y, c = x, d = z$, a zero-sum game if $a + w = b + x = c + y = d + z = 0$, a constant-sum game if $a + w = b + x = c + y = d + z = K \neq 0$ and a discoordination game if $c > a, b > d, w > x, z > y$ or $a > c, d > b, x > w, y > z$.

A systematic description of the outcomes in a game-theoretic problem is called a *solution.* Optimality and equilibria are the important solution concepts. In some games, there is an optimal way of taking an action independent of the actions of other players. A strategy is *Pareto optimal* (PO) if the only way each player has to increase his/her payoff is decreasing someone else's payoff. *Nash Equilibrium* (NE) is the most widely used solution concept in game theory. Nash Equilibrium is a position where all the strategies are mutually optimal responses. This means that no unilateral change of strategy will give a higher payoff to the corresponding player. Therefore, once it is reached, no individual has an incentive to deviate from it unilaterally. However NE concept, too, has difficulties: (i) Not every game has an NE in pure strategies, (ii) An NE need not be the best solution, or even a reasonable solution for a game. It is merely a stable solution against unilateral moves by a single player, and (iii) NE need not be unique. This prevents a sharp prediction with regard to the actual play of the game. In Fig.1, we give examples of classical games with different types of dilemmas.[1,20]

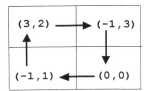 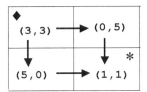 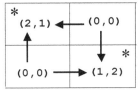

| **Samaritan□s Dilemma** | **Prisoners' Dilemma** | **Battle of Sexes** |

Fig. 1. Payoff matrices of games with different types of dilemmas: Samaritan's dilemma (SD) with no Nash equilibrium (NE), Prisoners' dilemma (PD) where NE and Pareto optimality (PO) do not coincide, and the Battle of Sexes (BoS) with two NE's. Horizontal (vertical) arrows correspond to the best responses of the column (row) player for a given strategy of the row (column) player. The entries pointed by two arrows and labelled with * are the Nash equilibria. The Pareto optimal entry in the PD game is labelled with ♦.

3. Basic Concepts of Quantum Mechanics for a Study of Games

In this section, we will introduce some basic concepts of quantum mechanical toolbox needed for the study of quantum versions of classical games. Reference[21] is an excellent resource for a detailed study of quantum mechanics and quantum information science.

3.1. *Quantum bits*

The state of a quantum system is described by a vector in Hilbert space. The simplest quantum mechanical system has a two dimensional state space, and can be represented as $|\psi\rangle = \alpha|0\rangle + \beta|1\rangle$ where $|0\rangle$ and $|1\rangle$ are basis states (vectors), and α and β are complex numbers satisfying $|\alpha|^2 + |\beta|^2 = 1$. This simplest system is called as *quantum bit* or *qubit*. While the classical bits can be in one of two states 0 or 1 which can in principle be distinguished, a qubit is in a continuum of states described by α and β and it is not possible, in general, to reliably distinguish non-orthogonal states of a quantum system. If a measurement is performed on the qubit, the outcome will be $|0\rangle$ with probability $|\alpha|^2$, and it will be $|1\rangle$ with probability $|\beta|^2$. A discussion of this measurement is given in subsection 3.4. Note that the normalization ensures that some result is obtained with probability 1.

3.2. *Density matrices*

This is a frequently used formalism in quantum mechanics. Suppose we have an ensemble of possible states $\{|\psi_1\rangle, |\psi_2\rangle, \ldots, |\psi_n\rangle\}$ with probabilities p_i.

This ensemble is a mixture of all $|\psi_i\rangle$'s and its density matrix is defined as

$$\hat{\rho} = \sum_{i=1}^{n} p_i |\psi_i\rangle\langle\psi_i|. \tag{1}$$

It is important to notice that this is a mixture but not a superposition of $|\psi_i\rangle$'s, but each of $|\psi_i\rangle$'s may be a superposition state. A density matrix may be formed from many different ensembles, e.g., the ensemble of $|0\rangle$ and $|1\rangle$ with equal probabilities and the ensemble of the superposition states $(|0\rangle + |1\rangle)/\sqrt{2}$ and $(|0\rangle - |1\rangle)/\sqrt{2}$ with equal probabilities have the same density matrix. No measurement can distinguish these two ensembles if no additional information is given about the preparation.

3.3. Unitary transformation

The evolution of an isolated quantum system is governed by unitary operator $\hat{U}$ which takes the state $|\psi\rangle$ at time $t = t_0$ to the state $|\psi'\rangle$ at time $t = t_1$. This evolution can be represented as $|\psi'\rangle = \hat{U}|\psi\rangle$. The evolution of a density matrix $\hat{\rho}$ is described by $\hat{\rho}' = \hat{U}\hat{\rho}\hat{U}^\dagger$.

A unitary operator on a qubit can be written as

$$\hat{U} = e^{i\phi} \begin{pmatrix} e^{i\gamma}\cos(\frac{\theta}{2}) & e^{i\delta}\sin(\frac{\theta}{2}) \\ -e^{-i\delta}\sin(\frac{\theta}{2}) & e^{-i\gamma}\cos(\frac{\theta}{2}) \end{pmatrix}. \tag{2}$$

Unitary operators without the global phase $e^{i\phi}$ belong to SU(2) operators set which will be mainly used in this study. The operators known as Pauli operators play an important role in quantum information, computation and games. These operators are

$$\hat{\sigma}_0 = \begin{pmatrix} 1 & 0 \\ 0 & 1 \end{pmatrix}, \ \hat{\sigma}_x = \begin{pmatrix} 0 & 1 \\ 1 & 0 \end{pmatrix}, \ \hat{\sigma}_y = \begin{pmatrix} 0 & -i \\ i & 0 \end{pmatrix}, \ \hat{\sigma}_z = \begin{pmatrix} 1 & 0 \\ 0 & -1 \end{pmatrix}. \tag{3}$$

The identity operator $\hat{I} = \hat{\sigma}_0$ leaves the qubits invariant. The bit flip operator $\hat{\sigma}_x$ transforms $|0\rangle$ to $|1\rangle$, and $|1\rangle$ to $|0\rangle$. $\hat{\sigma}_z$ is the phase flip operator which takes $|1\rangle$ to $-|1\rangle$ and does nothing on $|0\rangle$. The $\hat{\sigma}_y$ Pauli operator performs both a phase flip and a bit flip which takes $|0\rangle$ to $i|1\rangle$ and $|1\rangle$ to $-i|0\rangle$. Another important operator that we will frequently use is the Hadamard operator

$$\hat{H} = \frac{1}{\sqrt{2}} \begin{pmatrix} 1 & 1 \\ 1 & -1 \end{pmatrix}, \tag{4}$$

which takes $|0\rangle$ to $(|0\rangle + |1\rangle)/\sqrt{2}$ and $|1\rangle$ to $(|0\rangle - |1\rangle)/\sqrt{2}$.

3.4. *Measurement*

In order to gather information from a quantum mechanical system we have to perform measurement. Quantum measurements are described by a set of measurement operators $\{M_i\}$ acting on the state space to be measured. The index i is associated with the measurement outcome . If the state is $|\psi\rangle$ before measurement, then the post measurement state is

$$|\psi_i\rangle = \frac{M_i|\psi\rangle}{\sqrt{\langle\psi|M_i^\dagger M_i|\psi\rangle}} \tag{5}$$

with the measurement result i occuring with probability $p_i = \langle\psi|M_i^\dagger M_i|\psi\rangle$. The measurement operators satisfy the completeness equation $\sum_i M_i^\dagger M_i = \hat{I}$. As an example let us measure the qubit $|\psi\rangle = \alpha|0\rangle + \beta|1\rangle$ in the computational basis which is described by two measurement operators $M_0 = |0\rangle\langle0|$ and $M_1 = |1\rangle\langle1|$. First, observe that the measurement operators are Hermitian, $M_0 = M_0^\dagger$ and $M_1 = M_1^\dagger$, and they satisfy the completeness equation $M_0^\dagger M_0 + M_1^\dagger M_1 = M_0^2 + M_1^2 = M_0 + M_1 = \hat{I}$. Then the probability of obtaining zero as the measurement outcome is $p_0 = \langle\psi|M_0^\dagger M_0|\psi\rangle = \langle\psi|M_0|\psi\rangle = |\alpha|^2$. In the same way, the probability of obtaining the measurement outcome is $p_1 = \langle\psi|M_1^\dagger M_1|\psi\rangle = \langle\psi|M_1|\psi\rangle = |\beta|^2$, just as given in subsection 3.1. The post measurement states then are the $|0\rangle$ for M_0 and $|1\rangle$ for M_1.

A special class of the general measurement described above is the projective measurements. A projective measurement is defined by an observable which is a hermitian operator M acting on the state space to be measured. Upon measuring the system $|\psi\rangle$ with M, we obtain an outcome m_i which is an eigenvalue of M. If the projector onto the eigenspace of M with eigenvalue m_i is P_i, then the observable has a spectral decomposition $M = \sum_i m_i P_i$. The probability of getting the result m_i is given by $p_i = \langle\psi|P_i|\psi\rangle$. Then the post measurement state becomes $|\psi_i\rangle = P_i|\psi\rangle/\sqrt{p_i}$.

For some applications, we are not very much interested in the post-measurement state but rather on the probabilities of the respective measurement outcomes. For such situations, a mathematical tool called as POVM (positive operator valued measure) is used. POVM is a collection of positive operators $\{E_i\}$ with the completeness condition $\sum_i E_i = \hat{I}$. The probability of the measurement outcome is then given by $p_i = Tr[|\psi\rangle\langle\psi|E_i]$.

3.5. Correlations in quantum mechanical systems

Analysis of quantum games presumes that the players initially share some kind of correlation but they are not allowed to have any type of communication. Thus, it is needed to define the correlations in a quantum mechanical system. We start by giving the mathematical representation of an *uncorrelated* state defined as $\hat{\rho}_u = \hat{\rho}_1 \otimes \hat{\rho}_2 \otimes \cdots \otimes \hat{\rho}_{n-1} \otimes \hat{\rho}_n$ where measurement of observables on respective subsystems always factorize. On the other hand, a *classically correlated* system is described as $\hat{\rho}_c = \sum_i p_i \hat{\rho}_1^{(i)} \otimes \cdots \otimes \hat{\rho}_n^{(i)}$ with $\sum_i p_i = 1$. Unlike the uncorrelated system, joint measurement of observables for such a system do not factorize.

In quantum mechanical system, there is another type of correlation which cannot be seen in classical systems. This is the quantum correlation or so-called *entanglement*. A state is entangled if it cannot be written as a convex combination of product states. An entangled state cannot be prepared by local interactions even if the preparers can communicate classically. In entangled systems, one cannot acquire all information on the system by accessing only some of the subsytems. For a bipartite system, there are four maximally entangled states (MES) which are defined as $|\Phi^{\pm}\rangle = (|00\rangle \pm |11\rangle)/\sqrt{2}$ and $|\Psi^{\pm}\rangle = (|01\rangle \pm |10\rangle)/\sqrt{2}$. Starting from one of these bipartite MES, any bipartite state can be prepared using local operations and classical communication (LOCC).

On the other hand, in the multipartite case ($N \geq 3$), it is known that there are inequivalent classes of states such as those represented by the N-particle W-states $|W_N\rangle = |N-1,1\rangle/\sqrt{N}$ and the N-particle Greenberger-Horne-Zeilinger (GHZ) states $|\text{GHZ}_N\rangle = (|N,0\rangle + |0,N\rangle)/\sqrt{2}$. Here $|N-k,k\rangle$ corresponds to a symmetric state with $N-k$ zeros and k ones, e.g., $|W_3\rangle = |2,1\rangle/\sqrt{3} = (|001\rangle + |010\rangle + |100\rangle)/\sqrt{3}$ and $|\text{GHZ}_3\rangle = (|3,0\rangle + |0,3\rangle)/\sqrt{2} = (|000\rangle + |111\rangle)/\sqrt{2}$. In addition to these classes, there is the class of cluster states $|\chi_N\rangle$ which includes $|\text{GHZ}_3\rangle$ but forms another inequivalent class for $N \geq 4$. Those inequivalent classes cannot be transformed into each other by LOCC. Although the total amount of entanglement contained in $|W_N\rangle$ is smaller than the other two classes, it has higher persistency of entanglement which is defined as the minimum number of local measurements to completely destroy entanglement. It has been shown that $|W_N\rangle$ is optimal in the amount of pairwise entanglement when $N-2$ particles are discarded,[22,23] and has a persistency of $N-1$, whereas GHZ and cluster states have one and $N/2$, respectively.[24]

The states $|\Psi^+\rangle$ and $|W_N\rangle$ are the members of a more general class of symmetric states called as Dicke states, $|N-m,m\rangle/\sqrt{{}_N C_m}$ where ${}_N C_m$

denotes the binomial coefficient.[25,26] One can easily see that GHZ states can be written as a superposition of two Dicke states $|N,0\rangle$ and $|0,N\rangle$.

4. A Brief Review of Quantum Game Theory: Models and Present Status

In this section, we will present a short historical background of quantum game theory, and introduce, in detail, two quantization schemes which have been the milestones in this area of research. These schemes are Meyer's scheme for coin tossing game[15] and the Eisert's scheme for Prisoners' dilemma (PD).[16] The transfer of the classical games into the quantum domain has been done using the resources allowed by the quantum mechanical toolbox which includes unitary operators, superposition states, entanglement and quantum measurement.

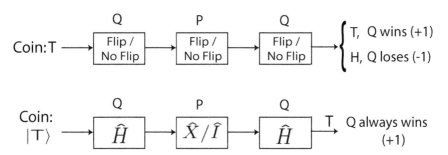

Fig. 2. Classical (upper) and quantum (lower) versions of the coin flip game. Coin is initially prepared in tails state (T). After the turns of P and Q, coin is analyzed. If it is tails (T), Q wins; otherwise Q loses. While in classical version strategy set players have the actions of Flip and No Flip, in the quantum version Q has the quantum strategy (operator) $\hat{H}$ while P has the classical strategies with the corresponding quantum operators $\{\hat{X}, \hat{I}\}$.

4.1. Coin flip game

The first strong argument for a quantum game theory was presented by Meyer who introduced the coin flip game between two Star Trek characters, Q and P.[15] In this simple game (see Fig. 2) P prepares a coin in the "tails" state and puts it into a box. Then Q can choose either to flip the coin or leave it unchanged without knowing the initial state of the coin. In the same way, P, without knowing the action of Q, can choose either of the actions. After the second turn of Q, the coin is examined. If the coin is tails, Q wins and if it is heads Q loses. A classical coin clearly gives both players an equal probability of success. As can be seen from the game payoff matrix in Table 2, there is no Nash equilibrium in the pure strategies, and the best the players can do is a mixed strategy.

In Meyers description of coin flip game, a quantum coin with states defined as $T = |0\rangle$ and $H = |1\rangle$ is used, and P is restricted to classical strategies $\{\text{Flip} = \hat{X}, \text{No Flip} = \hat{I}\}$ while Q has the quantum strategies $\{\hat{X}, \hat{H}\}$. Then the game sequences as follows

$$|0\rangle \xrightarrow[Q]{\hat{H}} 2^{-1/2}(|0\rangle + |1\rangle) \xrightarrow[P]{\hat{X} \text{ or } \hat{I}} 2^{-1/2}(|0\rangle + |1\rangle) \xrightarrow[Q]{\hat{H}} |0\rangle \qquad (6)$$

from which it is seen that just applying a Hadamard gate before and after P's action, Q brings the coin into the tail and hence wins the game regardless of the action of P. This work clearly shows that players with quantum strategies perform better than the players restricted to classical strategies.

Table 2.   Payoff matrix for the classical coin flip game where ‡ depicts the best strategy combination of Q for a given action of P.

	Q:NN	Q:NF	Q:FN	Q:FF
P: N	$(1, -1)$	$(-1, 1)$‡	$(-1, 1)$‡	$(1, -1)$
P: F	$(-1, 1)$‡	$(1, -1)$	$(1, -1)$	$(-1, 1)$‡

4.2. Eisert's model - Prisoners' Dilemma

In the same year as Meyer's work,[15] Eisert et al. studied a quantum version of PD where two players sharing an entangled state use quantum strategies.[16] Before going into the quantization of this game, let us first review the classical one: The so-called Prisoners' Dilemma game is an important and well-studied two player two strategy symmetric game where the players Alice and Bob independently without knowing the action of the other player decide whether to "cooperate" (C) or "defect" (D). According to their combined strategy choices, they obtain a certain payoff determined by the game payoff matrix (see Fig. 1 for the payoff matrix). They then face a dilemma since as rational players they should look for an equilibrium which dictates that they should both defect. On the other hand, they would

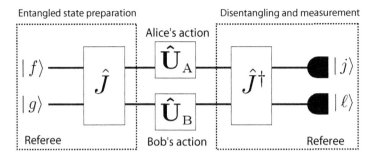

Fig. 3. Schematic configuration of Eisert's scheme to play two-player strategic games. The strategy set of the players are unitary operators $\hat{U}_A$ and $\hat{U}_B$, respectively for the players Alice and Bob. An entangled state is prepared by the referee using the entangling operator $\hat{J}$, and is distributed to players. After the action of the players referee performs measurement to distribute payoffs.

both benefit from cooperation which is the Pareto optimal. Their dilemma originates from the fact that NE and PO do not coincide.

In Eisert's model (see Fig.3), a two-player two-strategy game is played in the quantum mechanical settings as follows: (i) A referee prepares the maximally entangled state $|\Psi_i\rangle$ by applying the entangling operator $\hat{J}$ on a product state $|fg\rangle$ where

$$\hat{J}|fg\rangle = \frac{1}{\sqrt{2}}[\,|fg\rangle + i(-1)^{(f+g)}|(1-f)(1-g)\rangle\,]\qquad(7)$$

with $f, g = 0, 1$. Then the referee sends this state to two players, Alice and Bob, one qubit for each player. (ii) As their strategies, Alice and Bob carry out unitary operations $U_A, U_B \in SU(2)$ described by

$$\hat{U}(\theta, \phi, \psi) = \begin{pmatrix} e^{i\phi}\cos(\frac{\theta}{2}) & e^{i\psi}\sin(\frac{\theta}{2}) \\ -e^{-i\psi}\sin(\frac{\theta}{2}) & e^{-i\phi}\cos(\frac{\theta}{2}) \end{pmatrix}\qquad(8)$$

with $0 \leq \theta \leq \pi$, $0 \leq \phi \leq \pi/2$, and $0 \leq \psi \leq \pi/2$. They then send their qubits to the referees. (iii) After the referee receives the state, he operates $\hat{J}^\dagger$ obtaining $|\Psi_f\rangle = \hat{J}^\dagger(\hat{U}_A \otimes \hat{U}_B)\hat{J}|fg\rangle$. (iv) The referee makes a projective measurement $\Pi_n = |j\ell\rangle\langle j\ell|_{\{j,\ell=0,1\}}$ with $n = 2j + \ell$ corresponding to the projection onto the orthonormal basis $\{|00\rangle, |01\rangle, |10\rangle, |11\rangle\}$. In PD game, the measurement outcome $\{|00\rangle, |01\rangle, |10\rangle, |11\rangle\}$ corresponds to $\{(C,C), (C,D), (D,C), (D,D)\}$ where left and right entry represent Alice's and Bob's decision, respectively. (v) According to the measurement outcome n, the referee assigns to each player the payoff a_n and b_n chosen from the payoff matrix of the original classical game, e.g., $a_{\{n=0,1,2,3\}} = \{3, 0, 5, 1\}$

and $b_{\{n=0,1,2,3\}} = \{3,5,0,1\}$ in the PD game for Alice and Bob, respectively. Then the average payoff of the players can be written as

$$\$_A = \sum_n a_n \underbrace{\text{Tr}(\Pi_n \hat{J}^\dagger \hat{\rho}_f \hat{J})}_{P_{j\ell}},$$

$$\$_B = \sum_n b_n \underbrace{\text{Tr}(\Pi_n \hat{J}^\dagger \hat{\rho}_f \hat{J})}_{P_{j\ell}} \tag{9}$$

with $\hat{\rho}_f = |\Psi_f\rangle\langle\Psi_f|$ and $P_{j\ell}$ representing the probability of obtaining the measurement outcome n. When the referee starts with the input state $|fg\rangle = |00\rangle$, the probabilities $P_{j\ell}$ is found as

$$P_{00} = |x'\cos(\phi_A + \phi_B) - y'\sin(\varphi_A + \varphi_B)|^2$$
$$P_{01} = |x\sin(\phi_B - \varphi_A) - y\cos(\phi_A - \varphi_B)|^2$$
$$P_{10} = |x\cos(\phi_B - \varphi_A) - y\sin(\phi_A - \varphi_B)|^2$$
$$P_{11} = |x'\sin(\phi_A + \phi_B) + y'\cos(\varphi_A + \varphi_B)|^2$$
$$\tag{10}$$

where $x = \sin\frac{\theta_A}{2}\cos\frac{\theta_B}{2}$, $y = \sin\frac{\theta_B}{2}\cos\frac{\theta_A}{2}$, $x' = \cos\frac{\theta_A}{2}\cos\frac{\theta_B}{2}$, and $y' = \sin\frac{\theta_A}{2}\sin\frac{\theta_B}{2}$.

Note that original classical games should be simulated in a model of its quantum version, and in Eisert's model this is satisfied by imposing the following conditions:

$$[\hat{J}, \hat{\sigma}_y \otimes i\hat{\sigma}_y] = 0, \ [\hat{J}, \hat{\sigma}_0 \otimes i\hat{\sigma}_y] = 0, \ [\hat{J}, i\hat{\sigma}_y \otimes \hat{\sigma}_0] = 0 \tag{11}$$

These conditions imply that the unitary operator $\hat{U}(\theta,0,0) = \cos(\theta/2)\hat{\sigma}_0 + i\sin(\theta/2)\hat{\sigma}_y$ and $\hat{J}$ commute. Hence when players' strategies are restricted to $\hat{U}(\theta)$, the order of operations $\hat{J}$ and $\hat{U}_A \otimes \hat{U}_B$ can be changed and then $\hat{J}$ and $\hat{J}^\dagger$ are cancelled, resulting in the state $|\Psi_f\rangle = (\hat{U}_A \otimes \hat{U}_B)|00\rangle$. In this case, the joint probability $P_{j\ell}$ factorize $P_{j\ell} = p_A(j)p_B(\ell)$. For PD game $p(C) = \cos^2(\theta/2)$ and $p(D) = 1 - p(C)$, and the payoff function takes the form of the classical payoff function. This implies that $\hat{\sigma}_0$ and $i\hat{\sigma}_y$ correspond to the pure strategies C and D, and the setting can simulate the classical game faithfully. Therefore, when the players take $\hat{U}(\theta,0,0)$ as their strategies, the original Prisoners' dilemma is reproduced.

In this model, when players' strategy space is restricted to only one-parameter unitary operator $\hat{U}(\theta,0,0)$, the payoff functions become identical to those of the classical Prisoners Dilemma with mixed strategies. This equilibrium point gives 1 to each player which is less than the optimal one with 3 to each. On the other hand, when the players are allowed to use two-parameter unitary operators $\hat{U}(\theta, \phi, 0)$, a unique Nash equilibrium with payoff (3,3), which is also the Pareto optimal, emerges for their strategies $\hat{U}(\theta = 0, \phi = \pi/2, 0)$. Thus, it is claimed that by introducing quantum toolbox, the dilemma in this game disappears and players can achieve the Pareto optimal solution.

Benjamin and Hayden, however, criticized this restricted operation set to be unnatural and pointed out that if two players are allowed the full range of quantum strategies, there exists no Nash equilibrium in the game because the effect of one player's unitary transformation can be cancelled by the other locally using a counter-strategy due to the shared bipartite maximally entangled state.[27] This leads to a point that there is no compromised strategy. However, their argument does not hold when a non-maximally entangled state is shared between the players, and when the game is extended to multi-player form. Benjamin and Hayden have also examined three and four player quantum games which are strategically richer than the two player ones.[28] They constructed a three-player PD game and showed that the game has an NE even when the players are allowed to use full range of unitary operators. We should note that despite the critiques, the new dynamics brought into the games by quantum operators and entanglement in Eisert's model is clear though sometimes results are rather surprising.

4.3. Present status in quantum game theory

It is important to realize that quantization of a classical game is not unique; one can introduce quantumness into a game in different ways. The model proposed by Eisert *et al.* has been the widely accepted and studied quantization scheme despite some negative critiques as we have discussed above. Another way of achieving results similar to those of Eisert's scheme was introduced by Marinatto and Weber[29] who removed the disentangling operator $\hat{J}^\dagger$ and simply hypothesized various initial states. In this model, strategies are chosen as the probabilistic mixtures of identity and flip operators. Although the scheme is much simpler it does not allow probing the whole set of quantum strategies. Later, Nawaz and Toor[30] proposed a more general model which includes both the Eisert and Marinatto models. In this model, they introduced two entangling parameters (one for $\hat{J}$ and the other for $\hat{J}^\dagger$) which can be tuned. If the two parameters are equal we obtain the Eisert scheme, and when the parameter for $\hat{J}^\dagger$ is zero, we end up with the Marinatto schemes. Cheon and Tsutsui introduced a simple and general formulation of quantum game theory by accommodating all possible strategies in the Hilbert space, and solved two strategy quantum games.[31] Recently, Ichikawa *et al.* proposed a formulation of quantum game theory based on the Schmidt decomposition where they could quantify the entanglement of quantum strategies.[32]

The model proposed by Eisert *et al.* is a general one that can be applied to any two-player two-strategy game, with the generalization to N-player two-strategy games as well as two-player n-strategy games. While in the former, the players share multipartite entangled state and use SU(2) operators, in the latter they share bipartite entanglement but armed with SU(n) operators to represent the players' actions. Multiplayer games exhibit a richer dynamics than the two player games. While such games can be realized by preparing initially a multipartite entangled state, Chen *et al.* has proposed to play using only bipartite entangled states shared by neighboring players.[33]

In order to investigate the property of quantum version of dynamic games, Iqbal and Toor analyzed quantum version of repeated Prisoners' dilemma game where the players decide to cooperate in the first stage while knowing that both will defect in the second stage.[34] Kay *et al.* studied evolutionary quantum games and demonstrated that length of memory is crucial to obtain and maintain the quantum advantage.[35] Iqbal has also considered evolutionarily stable strategies (ESS) in quantum versions of both the prisoners dilemma and the battle of the sexes and concluded that entanglement can be made to produce or eliminate ESS's while retaining the same set of NE.[36] He also showed that it is only when entanglement is employed, a mutant strategy with quantum tools can easily invade a classical ESS. In these models the replicator dynamic takes a quantum form.[37,38]

Various quantum advantages over the classical have been found in the series of studies on games in quantum settings. Comparative studies by Shimamura *et al.*[20] and Ozdemir *et al.*[39] on the effects of classical/quantum correlations and operators not only confirmed the known effects of using the quantum mechanical toolbox in game theory but also revealed that in some circumstances one need not stick to quantum operations or correlations but rather use simple classical correlations and operations. Based on the requirement that classical game should be reproduced in its quantum model, they derived a necessary and sufficient condition for the entangled states and quantum operations which are allowed in playing games in quantum mechanical settings.[40,41]

Piotrowski and Sladkowski have proposed a quantum game theoretic approach to economics.[42–44] In their model of market, transactions are described as projective operations on Hilbert spaces of strategies of traders, and a quantum strategy is a superposition of trading actions and hence can achieve outcomes not realizable by classical means. They also speculated that markets will be more efficient with quantum algorithms and dramatic market reversals can be avoided.

There has been a growing interest in the quantum version of the well-known Parrondo's games,[13,45,46] where two games that are losing when played separately become a winning game when played in combination (their convex combination). It is expected that Parrondo games have the potential use in increasing the efficiencies of quantum algorithms. Flitney and Abbott have developed models and studied the history dependent quantum versions of these games.[47,48]

Physical realizations of Eisert's scheme of PD have been achieved first in a two qubit NMR computer[49] and recently in an all-optical one-way quantum computer.[50] The NMR experiment followed the original proposal by Eisert, and showed good agreement with theory. The amount of entanglement between the players was varied and the transition between classical and quantum regions were clearly observed. The all-optical realization was performed following the scheme of Paternostro et al.[51] This scheme is a hybrid one combining the quantum circuit approach and the cluster state model. The quantum circuit was realized by a 4-qubit box-cluster configuration, and the local strategies of the players by measurements performed on the physical qubits of the cluster.

Here, we gave only a short summary of what has been done in the field. Detailed and excellent reviews of quantum game theory can be found in the references.[40,52–54]

5. Study of a Discoordination Game - Samaritan's Dilemma - Using Eisert's Model: Effects of Shared Correlation on the Game Dynamics

Samaritan's Dilemma (SD), which is a discoordination game with no Nash equilibrium in classical pure strategies, is an interesting game where one can systematically analyze the changes in the game dynamics for various quantum mechanical settings. Games with no NE are interesting because they represent situations in which individual players might never settle down to a stable solution. Therefore, the important step in such games is to find an NE on which the players can settle down. In the following, we will investigate this game under different settings of Eisert's model. In particular we will see how the game payoff matrix and the strategies of the players evolve for different strategy sets and shared correlation between the players. A more detailed analysis can be found in References 39 and 40.

In the game of SD, the "Samaritan" (Alice) voluntarily wants to help the person "in need" (Bob) who cannot help himself. However, she suffers a welfare loss due to the selfish behavior of Bob who may influence or create situations which will evoke Alice's help. Then, a dilemma arises because Alice wants to help, however, Bob acts so strategically that the amount of help increases which is not desirable for Alice. Moreover, Alice cannot retaliate to minimize or stop this exploitation because doing so is a punishment for Bob and this will harm Alice's own interests in the short run.[55] The game and the payoff matrix (see Fig. 1) studied in this paper are taken from Ref.1 where the specific game is named as The Welfare Game. In this game, Alice (row player in the game matrix) wishes to aid Bob (column player in the matrix) if he searches for work but not otherwise. On the other hand, Bob searches for work if he cannot get aid from Alice. The action set of Alice is {Aid (A), No Aid (N)} whereas Bob's is {Work (W), Loaf (L)}. Thus their strategy combinations can be listed as (A, W), (A, L), (N, W), and (N, L).

When this game is played classically, there is no dominant strategy for neither of the players and there is no NE for pure strategies: (A,W) is not an NE because if Alice chooses A, Bob can respond with strategy L where he gets a better payoff, 3, as shown with arrow in Fig. 1. (A,L) is not an NE because, in this case, Alice will switch to N. The strategies (N,L) and (N,W) are not NE either, because for the former one Bob will switch to W to get payoff one, whereas for the latter case Alice will switch to A to increase her payoff from -1 to 3. Therefore, this game has no NE when played with pure classical strategies.

In mixed classical strategies, Alice randomize between her strategies A and N with the probabilities p and $(1 - p)$, respectively. In the same way, Bob's chooses between W and L with probabilities q and $(1 - q)$. Then the payoffs for Alice and Bob is written as

$$\$_A = 3pq - p(1 - q) - q(1 - p),$$
$$\$_B = 2pq + 3p(1 - q) + q(1 - p). \tag{12}$$

The strategy combination with $p = 0.5$ and $q = 0.2$ corresponds to the NE for the game with average payoffs given as $\$_A = -0.2$ and $\$_B = 1.5$. In this case, the payoff of Alice is negative which is not a desirable result for her. (A,L) and (N,L) emerge as the most probable strategies with probabilities 0.4.

In the study of this game, we first look at the problem whether a unique

NE exists for strategies utilizing classical or quantum mechanical toolboxes. If there exists at least one NE then we look at how the payoff of the players compare and to what extent this NE strategy can resolve the dilemma. For SD, there are three cases: CASE I, insufficient solution with $\$_A < 0$, CASE II, weak solution with $0 \leq \$_A \leq \$_B$, and CASE III, the strong solution solution with $0 \leq \$_B < \$_A$. If at the NE, both players receive positive payoffs then this implies both are satisfied with the outcome and there is no loss of resources for Alice (CASE II and III). The most desirable solution of the dilemma for Alice is represented in CASE III.

5.1. *Quantum operations and quantum correlations*

When the SD game is played with quantum operations but with no shared correlated state between the players, the situation produces the classical game with mixed strategies. Moreover, we know from Eisert's study that operators chosen from three parameter SU(2) set do not resolve the dilemmas in the games. Therefore, in the following we will not study these two cases. We will consider two types of MES: $|\psi_1\rangle = [|00\rangle + i|11\rangle]/\sqrt{2}$ and $|\psi_2\rangle = [|01\rangle - i|10\rangle]/\sqrt{2}$ obtained when the referee applies the entangling operator $\hat{J}$ on the initial states $|00\rangle$ and $|11\rangle$, respectively.

One-parameter SU(2) operators - Operators for Alice and Bob are obtained by setting $\phi_A = \varphi_A = 0$ and $\phi_B = \varphi_B = 0$ in Eq.(8). With the shared entangled state $|\psi_1\rangle$ and the one-parameter SU(2) operators, the payoffs for Alice and Bob become

$$\$_A = \frac{1}{4}[1 + 3(\cos\theta_A + \cos\theta_B) + 5\cos\theta_A\cos\theta_B],$$

$$\$_B = \frac{1}{2}[3 + 2\cos\theta_A - \cos\theta_A\cos\theta_B]. \tag{13}$$

From Eq.(13), we see that while $\$_B$ is always positive, $\$_A$ may be positive or negative depending on Bob's action. This implies that the samaritan, Alice, cannot make her payoff positive by acting unilaterally, thus Bob's choice of action becomes important to resolve Alice's dilemma. After some straightforward calculations, we find that an NE emerges for the operators $\hat{U}_A = (\hat{\sigma}_0 + i\hat{\sigma}_y)/\sqrt{2}$ and $\hat{U}_B = (\hat{\sigma}_0 + i2\hat{\sigma}_y)/\sqrt{5}$. At this unique NE, the payoffs of the players become $\$_A = -0.2$ and $\$_B = 1.5$. On the other hand, when they share $|\psi_2\rangle$, an NE is found for $\hat{U}_A = (\hat{\sigma}_0 + i\hat{\sigma}_y)/\sqrt{2}$ and $\hat{U}_B = (2\hat{\sigma}_0 + i\hat{\sigma}_y)/\sqrt{5}$ with the payoffs $\$_A = -0.2$ and $\$_B = 1.5$. It is seen that this strategy set reproduced the results of the classical mixed strategy, and does not bring any quantum advantage. This strategy set resolves the dilemma, but with an insufficient solution for Alice (CASE I).

Two-parameter SU(2) operators - This set of quantum operations is obtained from Eq. (8) by setting $\varphi_A = 0$ and $\varphi_B = 0$. For $|\psi_1\rangle$, the expressions for the payoffs are found as $\$_A = 3P_{00} - P_{01} - P_{10}$ and $\$_B = 2P_{00} + 3P_{01} + P_{10}$ with

$$P_{00} = \cos^2(\theta_A/2)\cos^2(\theta_B/2)\cos^2(\phi_A + \phi_B),$$
$$P_{01} = |x\sin\phi_B - y\cos\phi_A|^2,$$
$$P_{10} = |x\cos\phi_B - y\sin\phi_A|^2 \tag{14}$$

where $P_{j\ell}$ is calculated from Eq. (9). We find that there is a unique NE which appears at $\hat{U}_A = \hat{U}_B = i\hat{\sigma}_z$ with the payoff $(\$_A, \$_B) = (3,2)$. Note that this NE is a new one which cannot be seen in the classical version of the game. Since, this new NE emerges in pure quantum strategies we conclude that original discoordination game transforms into a coordination game. This is the NE where Alice receives the highest payoff available for her in this game, and where both players benefit. Therefore, Alice's dilemma is resolved in the stronger sense (CASE III).

In the case of $|\psi_2\rangle$, four NE's with equal payoffs $(\$_A, \$_B) = (3,2)$ emerge for the quantum operators $(\hat{U}_A, \hat{U}_B)$ equal to $(N, P) = (i\hat{\sigma}_y, i\hat{\sigma}_z)$, $(T, Q) = (\frac{\hat{\sigma}_0 + i\hat{\sigma}_y}{\sqrt{2}}, \frac{i(\hat{\sigma}_z + \hat{\sigma}_y)}{\sqrt{2}})$, $(Y, R) = (\frac{\hat{\sigma}_0 + i\sqrt{3}\hat{\sigma}_y}{2}, \frac{i(\sqrt{3}\hat{\sigma}_z + \hat{\sigma}_y)}{2})$, and $(Z, S) = (\gamma_0(\hat{\sigma}_0 + i\gamma_1\hat{\sigma}_y), i\delta_0(\hat{\sigma}_z + \delta_1\hat{\sigma}_y))$ where $\gamma_0 = \cos(3\pi/8)$, $\gamma_1 = \tan(3\pi/8)$, $\delta_0 = \cos(\pi/8)$, and $\delta_1 = \tan(\pi/8)$. At these NEs players receive payoffs higher than those when a classical mixed strategy is used. However, Alice's dilemma survives because players cannot coordinate their moves to decide on which NE point to choose. For example, if Alice thinks Bob will play $i\hat{\sigma}_z$ then she will play $i\hat{\sigma}_y$ to reach at the first NE. However, since this is a simultaneous move game and there is no classical communication between the players, Bob may play $i(\hat{\sigma}_z + \hat{\sigma}_y)/\sqrt{2}$ (because this action will take him to the second NE point) while Alice plays $i\hat{\sigma}_y$. Such a case will result in the case $\$_A < \$_B$ and will lower the payoffs of both players. Therefore, still a dilemma exists in the game; however the nature of dilemma has changed. In the classical version, the dilemma is due to welfare losses of Alice and lack of NE. However, now the dilemma is the existence of multiple NEs.

Implications - Now let us review what we have learned from the above analysis:

(1) Two-parameter SU(2) operators and shared entanglement are necessary to see the novel features introduced into game theory by the quantum mechanical toolbox. Note that one-parameter SU(2) operator set with shared entanglement corresponds to the classical mixed strategies.

(2) The number of solutions that is the number of NE's depends on the shared entangled state as well as on the payoff matrix of the original classical game. In SD game, switching from the entangled state $|\psi_1\rangle$ to $|\psi_2\rangle$ increases the number of NE from one to four. On the other hand, for the PD game, the number of NE's is one for both cases.

(3) The quantum mechanical toolbox does not necessarily resolve dilemma in classical games. Note that for the PD and SD played with $|\psi_1\rangle$ the dilemmas in the original games are resolved, however in SD played with $|\psi_2\rangle$, a new type of dilemma appears. While in the original game the dilemma was the absence of an NE where Alice is not exploited, in the quantum version played with $|\psi_2\rangle$ the dilemma is the existence of multiple NE's with the same payoffs. This new dilemma is similar to that of the Battle of Sexes (BoS) game, that is Alice and Bob cannot decide on which of the multiple NE's to choose.

(4) Quantum version of a classical game can be represented by an enlarged classical payoff matrix by including the unitary operators leading to equilibrium solutions into the payoff matrix of the initial classical game.[56] The originally 2×2 classical games of PD and SD are transformed into new games which can be described with the new 3×3 payoff matrices. In case of SD, another representation is the 5×6 payoff matrix obtained for $|\psi_2\rangle$. Evolution of the payoff matrix of the SD game under different correlated states is depicted in Fig. 4. Note that the new payoff matrix includes the payoff matrix of the original classical game as its submatrix.

(5) An important quantum advantage arises when we consider the communication cost of playing games. For a game represented by the 2×2 payoff matrix, each player needs 1 classical bit (c-bits) to classically communicate their preferred strategy to the referee. Thus, the total classical communication cost is 2 c-bits. When we play the quantum version of the game, the task is completed using four qubits (two prepared by the referee and distributed, and two qubits sent by the players back to the referee after being operated on) when shared entanglement is used. This corresponds to a communication cost of 2 e-bits (1 e-bit for the referee and 1 e-bit for the players). Now, if write down the payoff matrix for this quantum version, we end of with 3×3 payoff matrices for the PD and SD with $|\psi_1\rangle$, and a 5×6 matrix for SD with $|\psi_2\rangle$. The cost of communicating players' choices in these new payoff matrices classically (playing the game represented with those matrices) amounts to respectively to 4-cbits and 6-cbits. However, in quantum

versions what is needed is still 2 e-bits.

(6) In classical mixed strategies, the payoffs become continuous in the mixing probability of the players' actions, and a compromise becomes possible between the players which assures the existence of an NE. In games played by quantum operations, the inclusion of new strategies and moves into the game is an essential feature. Once these new moves are incorporated, both the strategy space and the payoffs become continuous. Then the players can finely adjust their strategies and come closer to finding the best responses to each other's actions which leads to one or more NE's.

5.2. Quantum operations and classical correlations

The shared entanglement introduced in the quantum version of the SD game transforms the original 2×2 non-cooperative game into a cooperative one. This new game thus has its own rules which are different than the original game: While in the original classical game, no communication of any form is allowed between the players, in the quantum version the players share entanglement which could be considered as a kind of "spooky communication".[57]

In order to clarify further the role of the operators and correlations in the obtained results above, we now consider that the players share classical correlations. This will make it easier to evaluate the power of quantum operators. Here, we consider that classical correlations are due to phase damping acting on the MES during its distribution (see Fig. 5). Phase damping transforms the initially prepared MES $|\psi_1\rangle$ and $|\psi_2\rangle$ into the classically correlated states $\hat{\rho}_{in,1} = [|00\rangle\langle 00| + |11\rangle\langle 11|]/2$ and $\hat{\rho}_{in,2} = [|01\rangle\langle 01| + |10\rangle\langle 10|]/2$, respectively. In the following, we will see how these correlations affect the games.

One-parameter SU(2) operators - It is straightforward to show that when the players are restricted to one-parameter SU(2) operators, they get the payoff $(0.25, 1.5)$ at the NE with $\hat{U}_A = \hat{U}_B = (\hat{\sigma}_0 + i\hat{\sigma}_y)/\sqrt{2}$ independent of the classical correlation they share. Alice's payoff in this case is much better than the one she receives when the players share MES and use one-parameter SU(2) operators.

Two-parameter SU(2) operators - In this new setting of the problem,

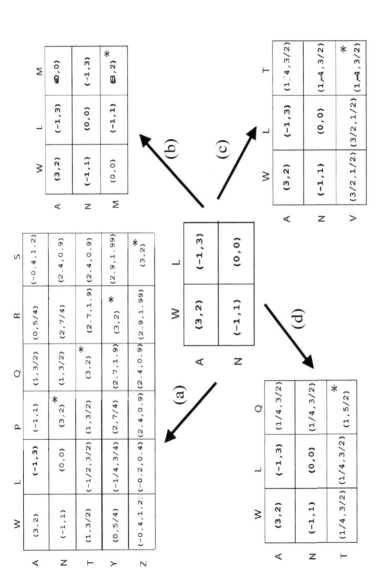

Fig. 4. Evolution of the payoff matrix of Samaritan's dilemma (SD) game for shared quantum and classical correlation between players using two-parameter SU(2) operators: (a) and (b) for quantum correlations $|\psi_2\rangle$ and $|\psi_1\rangle$, respectively, and (c) and (d) for classical correlations $\tilde{\rho}_{in,1}$ and $\tilde{\rho}_{in,2}$, respectively. These payoff matrices are obtained by adding to the original game matrix (2×2 payoff matrix in the middle) the strategies leading to Nash equilibria (NE). Refer to the text for the operators corresponding to the strategies shown by letters in these payoff matrices.

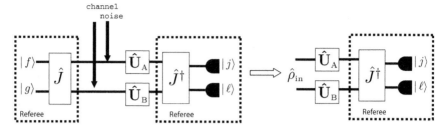

Fig. 5. Quantum version of a classical game played with classical correlations $\hat{\rho}_{in}$ obtained from an initial maximally entangled state (MES) after it is subjected to noise.

for $\hat{\rho}_{iii,1}$ the payoffs become

$$\$_A = \frac{1}{4}[1 + 5\cos\theta_A\cos\theta_B - 3\sin\theta_A\sin\theta_B\sin(\phi_A + \phi_B)],$$

$$\$_B = \frac{1}{2}[3 - \cos\theta_A\cos\theta_B - 2\sin\theta_A\sin\theta_B\cos\phi_A\sin\phi_B] \qquad (15)$$

from which we see that Alice cannot obtain a positive payoff by acting unilaterally so she should look for settlement at an equilibrium. It is easy to show that an NE with a payoff $(0.25, 1.5)$ appears for $\hat{U}_B = (\hat{\sigma}_0 + i\hat{\sigma}_y)/\sqrt{2}$ and $(\phi_A = 0$ and $\{\forall\theta_A : 0 \leq \theta_A \leq \pi\})$. This is a self-enforcing NE solution for the rational players Alice and Bob because neither of the players can make her/his payoff arbitrarily high and positive independent of the action of the other player. This NE provides a weak solution (CASE II).

For the shared classical correlation of $\hat{\rho}_{in,2}$, the the payoff functions are obtained from Eq. (15) by replacing all the $-$ by $+$ and vice verse except the $+$ in the $\sin(..)$ term. Then the operators $\hat{U}_A = (\hat{\sigma}_0 + i\hat{\sigma}_y)/\sqrt{2}$ and $\hat{U}_B = i(\hat{\sigma}_y + \hat{\sigma}_z)/\sqrt{2}$ leads to an NE with the payoff $(\$_A, \$_B) = (1, 2.5)$. This is a self-forcing NE, too, because Alice assures positive payoffs for both Bob and herself regardless of Bob's action. For this action of Alice, the rational Bob will go for the above operator which maximizes his payoff. Then, the dilemma is again resolved with a weak solution (CASE II).

Another possible correlation that may be shared by the players is the full rank classical correlation $\hat{\rho}_{in,3} = [|00\rangle\langle00| + |10\rangle\langle10| + |01\rangle\langle01| + |11\rangle\langle11|]/4$. In this case, regardless of their quantum operators players arrive at a weak solution (CASE II) with constant payoff $(0.25, 1.5)$.

Three-parameter SU(2) operators - It is only when the players share a MES that the three-parameter operator set does not give quantum advantage. It is then interesting to discuss whether this set of SU(2) operators affect the game dynamics in the case of shared classical correlation. For $\hat{\rho}_{in,1}$, the payoff functions can be obtained from Eq. (15) by replacing ϕ_A

and ϕ_B by $\phi_A + \varphi_A$ and $\phi_B + \varphi_B$, respectively. Then, an NE with payoffs ($\$_A = 1, \$_B = 2.5$) emerges for the operators ($\theta_A = \pi/2, \phi_A + \varphi_A = \pi$) and ($\theta_B = \pi/2, \phi_B + \varphi_B = \pi/2$). The extra parameter φ allows the players to increase their payoffs beyond those obtained with two-parameter SU(2) operators. On the other hand, for $\hat{\rho}_{in,2}$, we find that the strategies of the players to achieve an NE becomes ($\theta_A = \pi/2, \phi_A + \varphi_A = 0$) and ($\theta_B = \pi/2, \phi_B + \varphi_B = \pi/2$). These operators give them the payoffs ($\$_A = 1, \$_B = 2.5$) which are the same as they obtained with two-parameter operators. Note that even though the payoffs may increase, the solution is still a weak solution (CASE II).

Implications - Our findings in this section on the effects of classical correlations on the outcome can be summarized as follows:

(1) Entanglement is not necessary to find an NE that may solve the dilemma of the original game. Shared classical correlations between players with quantum operations leads to NE solutions which provide better payoffs for Alice compared to those she obtains for classical mixed strategies.
(2) Although a solution can be provided with the classical correlations considered in this study ($\hat{\rho}_{in,1}$, $\hat{\rho}_{in,2}$ and $\hat{\rho}_{in,3}$), the dilemma in SD is not resolved in the stronger sense (CASE III). The solution with the considered classical correlations is weak (CASE II). Note that with two-parameter SU(2) operators and shared MES $|\psi_2\rangle$, players obtain a strong solution.
(3) Three parameter SU(2) operators which cannot be used with shared MES can be utilized to find an NE with better payoffs. Note that the most cited criticism against quantum versions of classical games is that restricting the players to two-parameter SU(2) operators is not reasonable. Now it is possible to get rid of this. By decreasing the amount of shared entanglement, the players can exploit whole SU(2) without any restriction. It is important to keep in mind that this conclusion strongly depends on the payoff matrix and the dilemma type of the original classical game.

5.3. *Classical operations and classical correlations*

Classical operations are a subset of quantum operations, and in Eisert model they should commute with the entangling operator $\hat{J}$. For the SD game, quantum operators corresponding to classical strategies are the identity operator $\hat{\sigma}_0$ for A and W, and the flip operator $i\hat{\sigma}_y$ for N and L . Here

Table 3. The new payoff matrix for the SD game when players share the classical correlation $\hat{\rho}_{in} = [|00\rangle\langle00| + |11\rangle\langle11|]/2$, and use classical operations.

	Bob: $\hat{\sigma}_0$		Bob: $i\hat{\sigma}_y$
Alice: $\hat{\sigma}_0$	$(1.5, 1)$	$\rightarrow$	$(-1, 2)$
	$\uparrow$		$\downarrow$
Alice: $i\hat{\sigma}_y$	$(-1, 2)$	$\leftarrow$	$(1.5, 1)$

we consider the classical correlations as in the above subsection. When the classical correlation is $\hat{\rho}_{in,1}$, the diagonal and off-diagonal elements of the payoff matrix of the original game (see Fig. 1) are averaged out separately, resulting in a new game with the payoff matrix given in Table 3 without any correlation. From Table 3, it is seen that the game is still a discoordination game with no NE and the dilemma survives. When the classical correlation is $\hat{\rho}_{in,2}$, the payoff matrix has the same properties as in Table 3 with diagonal and off-diagonal elements interchanged. Thus, we conclude that classical correlation does not change the structure of the SD game except the scaling of the payoffs when players choose to play in classical pure strategies. However, when they switch to mixed strategies, that is they randomize between their pure strategies, we find that there is a unique NE when $p = 0.5$ and $q = 0.2$ with payoff $(0.25, 1.5)$. The values of p and q are the same as those in the classical mixed strategy of the original game without any shared correlation. But since the payoffs were scaled with the introduction of the classical correlation, now we see that there is an increase in Alice's payoff from -0.2 to 0.25. This is interesting because an NE appears where both Alice and Bob have positive payoff, implying that the dilemma is resolved in the weaker sense (CASE II). Note that this was not possible for the classical game played without correlations.

5.4. Classical operations and quantum correlations

As pointed out above, in Eisert's scheme, the classical operators of the players should commute with the entangling operators. This requirement reduces the game using classical operators to the case where players apply their operators to the initial product state from which the referee prepares the maximally entangled state, and then makes a projective measurement onto the four basis $\{|00\rangle, |01\rangle, |10\rangle, |11\rangle\}$ to calculate their payoff. This is the same as the original classical game played in pure and mixed strategies.

5.5. Classical Bob versus quantum Alice

The dilemma in SD game is a one-sided dilemma of only Alice, thus one may wonder whether Alice can do better and resolve her dilemma if Bob is restricted to classical strategies, and Alice is allowed to use quantum operations from SU(2) set when they share the MES, $|\psi_1\rangle$. In this case, Bob's operators will be $\hat{U}_B^0 = \hat{\sigma}_0$ with probability p and $\hat{U}_B^1 = i\hat{\sigma}_y$ with probability $1 - p$. The state after the actions of Alice and Bob becomes

$$\hat{\rho}_{\text{out}} = \sum_{k=0}^{1} p_k (\hat{U}_A \otimes \hat{U}_B^k) \hat{\rho}_{\text{in}} (\hat{U}_A^\dagger \otimes \hat{U}_B^{k\dagger}). \tag{16}$$

Then average payoffs are calculated as

$$\$_A = \sum_n a_n (pP'_{j\ell} + (1-p)P''_{j\ell}),$$

$$\$_B = \sum_n b_n (pP'_{j\ell} + (1-p)P''_{j\ell}). \tag{17}$$

where

$$P_{00} = p\cos^2(\frac{\theta_A}{2})\cos^2\phi_A + (1-p)\sin^2(\frac{\theta_A}{2})\sin^2\varphi_A,$$

$$P_{01} = (1-p)\cos^2(\frac{\theta_A}{2})\cos^2\phi_A + p\sin^2(\frac{\theta_A}{2})\sin^2\varphi_A,$$

$$P_{10} = (1-p)\cos^2(\frac{\theta_A}{2})\sin^2\phi_A + p\sin^2(\frac{\theta_A}{2})\cos^2\varphi_A. \tag{18}$$

One-parameter SU(2) operators - Setting $\phi_A = \varphi_A = 0$, an NE with payoffs $(\$_A, \$_B) = (-0.2, 1.5)$ emerges when Alice chooses $\theta_A = \pi/2$ corresponding to $\hat{U}_A = (\hat{\sigma}_0 + i\hat{\sigma}_y)/\sqrt{2}$ and Bob chooses $p = 0.2$. This is the same result as the case for quantum Alice and Bob with one-parameter SU(2) operators. Therefore, results of classical mixed strategies are reproduced. This is not very desirable for Alice due to her negative payoff while Bob's is positive (CASE I).

Two-parameter SU(2) operators - Setting $\varphi_A = 0$, we see that Alice can never achieve a positive payoff independent of Bob's strategy defined by p. For any strategy of Alice, Bob can find a p value that will minimize Alice's payoff and vice verse. This leads to the fact that there is no NE where both players can settle.

Three-parameter SU(2) operators - The situation changes dramatically in this case because Alice now has an extra parameter to control the outcome. When she chooses $\hat{U}_A = [\hat{\sigma}_0 + i(\hat{\sigma}_x + \hat{\sigma}_y + \hat{\sigma}_z)]/2$ or $\hat{U}_A = (\hat{\sigma}_0 + i\hat{\sigma}_x)/\sqrt{2}$, the payoffs become independent of Bob's action. Thus Alice can control the game but this does not let her to resolve the dilemma with a strong solution. The best she can do is a weaker solution (case II) with the corresponding payoffs $(0.25, 1.5)$ and $(1, 2.5)$.

It is thus understood that by acting unilaterally Alice can resolve her dilemma only when she is allowed to use operators chosen from three-parameter SU(2) set and Bob is restricted to classical operations. Even in this case, only a solution satisfying CASE II is achieved.

6. Decoherence and Quantum Version of Classical Games

It is well-known that quantum systems are easily affected by their environment, and physical schemes are usually far from ideal in practical situations. Therefore, it is important to study whether the advantage of players and dynamics of games with quantum operations and entanglement survive in the presence of noise and non-ideal components. In the previous sections, we assumed that the sources used in the realization of the games were ideal and free of noise but the distribution channel induced phase damping.

Since the initial state from which the referee prepares the entangled state is a crucial parameter in Eisert's scheme, any deviation from the ideality of the source will change the dynamics and outcomes of the game. Consequently, the analysis of situations where the source is corrupt is necessary to shed a light in understanding the game dynamics in the presence of imperfections. In this section, we consider a corrupt source and analyze its effects:[59] (i) Is there a critical corruption rate above which the players cannot maintain their quantum advantage if they are unaware of the action of the noise on the source, and (ii) How can the players adopt their actions if they have information on the corruption rate of the source. The first of such studies was carried out by Johnson[58] who has considered a three-player PD game and shown that a crossover from quantum to classical advantage occurs as the corruption rate increases.

Here, we consider a more general corrupt source model (see Fig. 6), where the initial state input to the entangler in Eisert's model is flipped with some probability, and study its effect on PD and SD games.[59] In the comparison of the classical and quantum versions with and without corruption in the sources, we assume that in the quantum version the players choose their optimal operators from the two-parameter SU(2) operators set

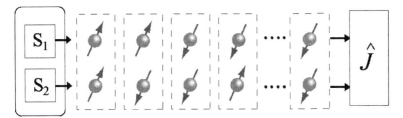

Fig. 6. Model of the corrupt source. Two identical sources S_1 and S_2 prepare the initial product state $|f\rangle|g\rangle$. In ideal case, they deterministically prepare the spin-down state $|0\rangle$ resulting in $|f\rangle|g\rangle = |0\rangle|0\rangle$. The corrupt sources, however, prepare the spin-up and -down states $|1\rangle$ and $|0\rangle$ with probabilities r and $1-r$, respectively. Therefore, the input state of the entangler, denoted by $\hat{J}$, becomes a mixture of spin-up and spin-down states.

while in the classical version they apply either $\hat{\sigma}_0$ or $i\hat{\sigma}_y$.

The state prepared by the corrupt source model shown in Fig. 6 can be written as $\hat{\rho}_1 = \hat{\rho}_2 = (1-r)|0\rangle\langle 0| + r|1\rangle\langle 1|$. Then the combined state generated and sent to the entangler becomes $\hat{\rho}_1 \otimes \hat{\rho}_2 = (1-r)^2|00\rangle\langle 00| + r^2|11\rangle\langle 11| + r(1-r)(|01\rangle\langle 01| + |10\rangle\langle 10|)$. This results in a mixture of the four possible maximally entangled states $(1-r)^2|\psi^+\rangle\langle\psi^+| + r^2|\psi^-\rangle\langle\psi^-| + r(1-r)(|\phi^+\rangle\langle\phi^+| + |\phi^-\rangle\langle\phi^-|)$, where $|\psi^\mp\rangle = |00\rangle \mp i|11\rangle$ and $|\phi^\mp\rangle = |01\rangle \mp i|10\rangle$. This is the state on which the players will perform their unitary operators.

The players assume ideal source while it is not - In this case, Alice and Bob are not aware of the corruption in the source, and they apply the optimal strategies to resolve their dilemma based on the assumption that the source is ideal and the referee distributes the MES $|\psi^+\rangle$.

For PD with ideal sources, we know that the best strategy of the players is to apply $\hat{U}_A = \hat{U}_B = i\hat{\sigma}_y$ in the classical version, and $\hat{U}_A = \hat{U}_B = i\hat{\sigma}_z$ in the quantum version. Then a quantum advantage is observed in the payoffs and dynamics of the game: an NE coinciding with the Pareto optimal is obtained. In quantum version, players receive \$$_A$ = \$$_B$ = 3 which is better than the payoff \$$_A$ = \$$_B$ = 1 in classical version. In the corrupt source model, we see that with increasing corruption rate r the payoffs in the quantum version decrease while those in the classical version increase. Then the payoffs with classical and quantum strategies become equal to 2.25 when $r = r^\star = 1/2$. If r satisfies $0 \le r < 1/2$, the quantum version of the game always does better than the classical one. Otherwise, the classical game is better. Thus, if $r > r^\star$, then the players would rather apply their classical strategies than the quantum ones. This can be explained as follows: Since classical operators commute with the entangling operator, in the classical version the game is played with no entanglement. Then the players apply

$i\hat{\sigma}_y$ directly on the state prepared by the source. If $r = 0$, then they operate on $|00\rangle$ and receive $\$_A = \$_B = 1$. On the other hand, if $r = 1$, they operate on $|11\rangle$ and receive $\$_A = \$_B = 3$. Thus, when the players apply the classical operator $i\hat{\sigma}_y$, their payoffs continuously increase from one to three with the increasing corruption rate from $r = 0$ to $r = 1$. In the quantum version, on the other hand, when $r = 0$, the optimal strategy is $\hat{U}_A = \hat{U}_B = i\hat{\sigma}_z$; however it will not be optimal when $r = 1$. The optimal strategy in this case is when $\theta'_A = \theta'_B = 0$ and $\phi'_A = \phi'_B = \pi/4$. If the players insist applying $i\hat{\sigma}_z$ when $r = 1$ then they will end up with the payoffs of the classical version. The results of the analysis are depicted in Fig.7.

In SD, we know that the best strategy for the players in the classical version is to use a mixed strategy which delivers $(\$_A, \$_B) = (-0.2, 1.5)$ at the NE. In this strategy, while Alice randomizes between her operators with equal probabilities, Bob uses a biased randomization and applies $\hat{\sigma}_0$ with probability 0.2. In the quantum version with ideal sources, we know that the most desired solution $(\$_A, \$_B) = (3, 2)$ is obtained at the NE when both players apply $i\hat{\sigma}_z$ to $|\psi^+\rangle$. The dynamics of the payoffs in this game with the corrupt source when the players stick to their operators $i\hat{\sigma}_z$ and its comparison with their classical mixed strategy are depicted in Fig. 7. In this asymmetric game the payoffs of the players become equal for the corrupt sources when $r = 1/7$ and $r = 1$. For these corrupt sources, they receive the payoffs 96/49 and 0, respectively. It is observed that the transition from the quantum advantage to classical advantage occurs at $r^\star = 1/2$ for both players. While for increasing r, $\$_B$ monotonously decreases from two to zero, $\$_A$ reaches its minimum of -0.2 at $r = 0.8$, where it starts increasing to the value of zero at $r = 1$. It is worth noting that when the players apply their classical mixed strategies in this physical scheme, $\$_B$ is always constant and independent of the corruption rate, whereas $\$_A$ increases linearly for $0 \leq r \leq 1$. We identify three regions of r corresponding to the three classes of solution to SD game: CASE I when $0.6 \leq r \leq 1$, CASE II when $1/7 \leq r < 0.6$, and finally CASE III for $r < 1/7$.

Players know the corruption in the source - We assume that the referee knows the characteristics of the corruption in the source, and informs the players on r. The question now is whether given r the players can find a unique NE; and if they can, does this NE resolve their dilemma in the game or not. One can easily see that for $r = 1/2$, the state shared between the players become $\hat{\rho} = \hat{I}/4$. Then independent of what action they choose, the players receive constant payoffs determined by averaging the payoff entries of the game matrix. For the symmetric game PD, both players receive the

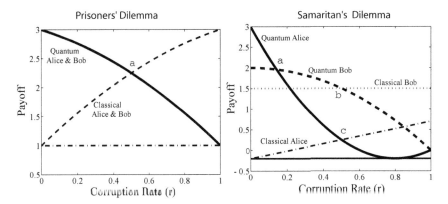

Fig. 7. The payoffs received by the players in Prisoners' dilemma (PD) and Samaritan's dilemma (SD) as a function of corruption rate, r in the source, for their quantum, and classical strategies. In the PD, the point labelled as a at $r = 1/2$ is the transition from quantum advantage to classical advantage. The horizontal dashed line denotes the payoffs of both players for the classical strategy without corruption. In SD, the labelled points are as follows: a, transition point from $\$_A > \$_B$ to $\$_B > \$_A$ (Case III $\leftrightarrow$ Case II), b and c at $r = 1/2$, transitions from quantum advantage to classical advantage for Bob and Alice, respectively. The horizontal solid line denotes the payoff Alice receives in classical strategies when the source is ideal. For Bob, classical strategies with and without noise coincide and are depicted with the horizontal dotted line.

payoff $9/4$; for the asymmetric game SD the payoffs become $1/4$ and $3/2$ respectively for Alice and Bob. We see from Table 4 that if the players know r then they can solve the dilemma for both $r = 0$ and $r = 1$ obtaining the payoffs 3. For the completely corrupt source $r = 1$, they should choose $\hat{U}_A = \hat{U}_B = (\hat{\sigma}_0 + i\hat{\sigma}_z)/\sqrt{2}$. This is however not true for the SD game: for $r = 0$ the dilemma is resolved with the best possible solution (CASE III) contrary to the situation for $r = 1$ where there is no unique NE (see Table 5). There are a number of NE's with equal payoffs which prevents players making sharp decisions.

Implications - The study on the corrupt source model reveals that:

(1) In a game with corrupt source with no side information to the players, quantum advantage no longer survives if the corruption rate is above a critical value.

(2) The strength of corruption may not only decrease the payoffs but also may lead to multiple NE's depending on the game even if the players know the corruption in the source.

(3) If the players are given the corruption rate in the source, they can either adapt to new optimal operators or choose a risk-free action, i.e. For PD

Table 4. Strategies leading to NE and the corresponding payoffs in the PD game for the source corruption rate r.

	$r = 0$	$r = 1/4$	$r = 3/4$	$r = 1$
$\hat{U}_A(\theta_A, \phi_A)$	$(0, \frac{\pi}{2})$	$(0, \frac{\pi}{2})$	$(0, \frac{\pi}{4})$	$(0, \frac{\pi}{4})$
$\hat{U}_B(\theta_B, \phi_B)$	$(0, \frac{\pi}{2})$	$(0, \frac{\pi}{2})$	$(0, \frac{\pi}{4})$	$(0, \frac{\pi}{4})$
$(\$_A, \$_B)$	$(3, 3)$	$(\frac{43}{16}, \frac{43}{16})$	$(\frac{43}{16}, \frac{43}{16})$	$(3, 3)$

Table 5. Strategies leading to NE and the corresponding payoffs in the SD game for the source corruption rate r. ^{1} $\forall \phi_A, \forall \phi_B \in [0, \pi/2]$ and ^{2} $\phi \in [0, \pi/4]$

	$r = 0$	$r = 1/4$	$r = 3/4$	$r = 1$
$\hat{U}_A(\theta_A, \phi_A)$	$(0, \frac{\pi}{2})$	$(0, \frac{\pi}{2})$	$(0, \phi)^1$	$(0, \phi)^2$
$\hat{U}_B(\theta_B, \phi_B)$	$(0, \frac{\pi}{2})$	$(0, \frac{\pi}{2})$	$(0, \frac{\pi}{2} - \phi)^1$	$(0, \frac{\pi}{2} - \phi)^2$
$(\$_A, \$_B)$	$(3, 2)$	$(\frac{21}{16}, \frac{15}{8})$	$(\frac{21}{16}, \frac{15}{8})$	$(3, 2)$

and SD games, they may choose the operators $\hat{U}_A = \hat{U}_B = (\hat{\sigma}_0 + i\hat{\sigma}_y)/\sqrt{2}$ where the payoffs of the players become constant independent of corruption rate.

(4) Although the quantum players have high potential gains, there is a large potential loss if the source is deviated from the ideal one. Classical players are more robust to corruption in the source.

It is worth to note here that with a decoherence model in the distribution channels in various two player quantum games in Eisert's model, Flitney *et al.* showed that the quantum player maintains an advantage over a player with classical strategies provided that some level of coherence remains.[60] In a recent study, Flitney and Hollenberg studied the Minority game in the presence of decoherence, and showed that the NE payoff is reduced as the decoherence is increased while the NE remains the same provided the decoherence probability is less than $1/2$.[61] The decrease rate in payoffs is much higher as the number of players in the game increases. They also showed that all players are equally disadvantaged by decoherence in one of the qubits. Hence no player, or group of players, can gain an advantage over the remainder by utilizing quantum error correction to reduce the error probability of their qubit.

7. Entanglement and Reproducibility of Multiparty Classical Games in Quantum Mechanical Settings

The expectation that quantum game theory may benefit from the rich dynamics of physical systems with many interacting particles ignited the research of multi-player games. This was also encouraged by the observation that entanglement leads to equilibrium solutions which are not observed in the classical versions of the games.

The first study of multiplayer quantum games was carried out by Benjamin and Hayden who showed by studying the PD and the Minority game that multiplayer games can exhibit forms of pure quantum equilibrium that can be observed neither in the classical versions nor in their two-player forms.[28] They also concluded that shared entanglement leads to different cooperative behavior and acts a kind of contract preventing players betraying each other. In their model, they considered GHZ-like states. Following this work, Y-J Han *et al.* performed a comparative study of GHZ and W-states in a three player PD game showing that (i) players prefer to form three-person coalition when the game is played with W state, and as a result they receive the highest payoff, and (ii) when the game is played with GHZ state, players prefer two-person coalitions and the players in the coalition receive payoff better than the third player who is not in the coalition.[62] They further compared their results with the classical counterparts to show the advantage of quantum games over the classical ones.

With the above two studies, it became generally accepted that quantum versions can be easily extended to N-player situations by simply allowing N-partite entangled states. In this section, however, we show that this is not generally true because the reproducibility of classical tasks in quantum domain imposes limitations on the type of entanglement and quantum operators. While in the case of GHZ state, the original classical game payoff matrix in pure strategies becomes a subset of the quantum version, this is not true for W-states. When the operators corresponding to the classical strategies are applied, the payoffs delivered to the players are unique entries from the classical payoff matrix for the GHZ state. However, for W-state the payoffs to the players become a probability distribution over the entries of the payoff matrix. Therefore, comparing the games played with these two states with each other and with the classical games in terms of the payoffs is not fair. For two-strategy games, we derived the necessary and sufficient conditions for a physical realization.[41,63,64] We also gave examples of entangled states that can and cannot be used, and the characteristics of quantum operators used as strategies.

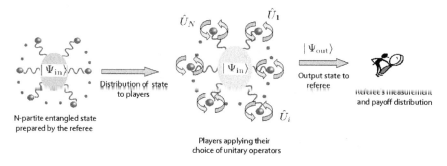

Fig. 8. The model for the multiplayer quantum games.

7.1. Reproducibility criterion to play games in quantum mechanical settings

In order to make a fair comparison between the classical and quantum versions of a game as well as its dynamics with different shared entangled states, the physical scheme in the quantum mechanical setting should be able to reproduce the results of the classical version, too. That is with the proper choices of operators the classical game should pop up as a subset of its quantum version. In Eisert's model of quantum games, reproducibility criterion is satisfied by choosing the quantum operators corresponding to classical strategies in such a way that they commute with the entangling operator $\hat{J}$. Therefore, we require that in any model of multiplayer games this criterion should be satisfied.

An N-player two-strategy classical game should be reproduced in the quantum version when each player's strategy set is restricted to two unitary operators, $\{\hat{u}_i^1, \hat{u}_i^2\}$, corresponding to the two pure strategies in the classical game. Then the combined pure strategy of the players is represented by $\hat{x}_k = \hat{u}_1^{l_1} \otimes \hat{u}_2^{l_2} \otimes \cdots \otimes \hat{u}_N^{l_N}$ with $l_i = \{1,2\}$ and $k = \sum_{i=1}^{N}(l_i - 1)2^{i-1}$. Thus the output state becomes $|\Phi_k\rangle = \hat{x}_k|\Psi\rangle$ with $k = \{1, 2, \ldots, 2^N\}$. For the strategy combination $\hat{x}_k$, expected payoff for the i-th player becomes $\$_i(\hat{u}_1, \cdots, \hat{u}_N) = \mathrm{Tr}\left[\left(\Sigma_j a_j^i \mathcal{P}_j\right) \hat{x}_k|\Psi\rangle\langle\Psi|\hat{x}_k^\dagger\right]$. In this section, we will look for an answer to the question: Can we find $\{\hat{u}_i^1, \hat{u}_i^2\}$ for any N-partite entangled state and for any game?

Reproducibility can be discussed in two cases: (i) The referee should be able to identify the strategy played by each player deterministically regardless of the structure of the payoff matrix, (ii) The referee should be able to reproduce the expected payoff of the classical game in the quantum version.[63] In (i), the referee should identify all possible outcomes $|\Phi_k\rangle$ deterministically which requires that all possible states formed by the joint applications of the players' operators should be mutually orthogonal to each other,

$$\langle \Phi_\alpha \mid \Phi_\beta \rangle = \delta_{\alpha\beta} \ \forall \alpha, \beta. \tag{19}$$

This strong criterion should be satisfied independent of the structure of the game payoff matrix. In (ii), however, the referee needs to distinguish between the subsets of outcomes with the same payoff. If two output states $|\Phi_j\rangle$ and $|\Phi_k\rangle$ have the same payoffs then they should be grouped into the same set. If all possible output states are grouped into sets $S_j = \{|\Phi_{1j}\rangle, |\Phi_{2j}\rangle, \ldots |\Phi_{nj}\rangle\}$ and $S_k = \{|\Phi_{1k}\rangle, |\Phi_{2k}\rangle, \ldots |\Phi_{n'k}\rangle\}$ then the referee should deterministically discriminate between these sets which is possible iff the state space spanned by the elements of each set are orthogonal. Hence for every element of S_j and S_k, we have $\langle \Phi_{nj}|\Phi_{n'k}\rangle = 0, \forall j \neq k$, that is all the elements of S_j and S_k must be orthogonal to each other, too. This weak criterion is very much dependent on the structure of the game payoff matrix.

When a two-player two-strategy game is extended to N-player game $(N > 2)$, the new payoff matrix is formed by summing the payoffs that each player would have received in simultaneously playing the two-player game with $N - 1$ players. Hence, in their N-player extensions, Prisoners' dilemma (PD), Samaritan's dilemma (SD), Boxed Pigs (BP), Modeller's dilemma (MD), Dead-Lock (DL), and Ranked Coordination (RC) have payoff matrices where all the outcomes have different payoff vectors.[1,41] Thus, these games should be evaluated according to the strong criterion. On the other hand Stag-Hunt (SH), Chicken Game (CG), Battle of Sexes (BoS), Battle of Bismarck (BB), Matching Pennies (MP), and Alphonse & Gaston Coordination Game (AG), Minority, majority and coordination games have payoff matrices where some of the outcomes have the same payoff vectors, thus these games can be evaluated by the weak criterion.[1,41]

7.2. Entangled states and strong reproducibility criterion

In this subsection, first as an example we will show that W-states cannot be used in quantum games because they do not satisfy the strong reproducibility criterion. Then we will derive a general form of entangled states that satisfy the criterion and hence can be used in quantum versions of classical games.

W states do not satisfy the strong reproducibility criterion - Let us assume that the N-partite W state $|W_N\rangle$ has been prepared and distributed to the players, and the i-th player has the strategy set $\{\hat{u}_i^1, \hat{u}_i^2\}$. Acting on the shared state $|W_N\rangle$ with all possible strategy combinations of N-players will generate the output state set $[|\Phi_1\rangle, |\Phi_2\rangle, ..., |\Phi_{2^N}\rangle)]$ such that

$$|\Phi_1\rangle = \left(\hat{u}_1^1 \otimes \hat{u}_2^1 \otimes \hat{u}_3^1 \otimes ... \otimes \hat{u}_N^1\right)|W_N\rangle$$

$$|\Phi_2\rangle = \left(\hat{u}_1^2 \otimes \hat{u}_2^1 \otimes \hat{u}_3^1 \otimes ... \otimes \hat{u}_N^1\right)|W_N\rangle$$

$$|\Phi_3\rangle = \left(\hat{u}_1^2 \otimes \hat{u}_2^2 \otimes \hat{u}_3^1 \otimes ... \otimes \hat{u}_N^1\right)|W_N\rangle$$

$$.. \qquad ....$$

$$.. \qquad ....$$

$$|\Phi_{2^N}\rangle = \left(\hat{u}_1^2 \otimes \hat{u}_2^2 \otimes \hat{u}_3^2 \otimes ... \otimes \hat{u}_N^2\right)|W_n\rangle. \tag{20}$$

Since the strong criterion of reproducibility requires that $\langle\Phi_\alpha|\Phi_\beta\rangle = \delta_{\alpha\beta}$ for $\forall\alpha, \beta \in \{1, .., 2^n\}$, we check the mutually orthogonality of all the 2^N output states listed above. We first take any two output states for which the strategies of only one player differ, e.g.,$|\Phi_1\rangle$ and $|\Phi_2\rangle$. The inner products of these states becomes

$$\langle\Phi_1 | \Phi_2\rangle = \frac{1}{N}[(N-1)\langle 0 |\hat{u}_1^{1\dagger}\hat{u}_1^2| 0\rangle + \langle 1 |\hat{u}_1^{1\dagger}\hat{u}_1^2| 1\rangle]. \tag{21}$$

Since $\hat{u}_1^1$ and $\hat{u}_1^2$ are SU(2) operators, so is the $\hat{u}_1^{1\dagger}\hat{u}_1^2$ which can be expressed as

$$\begin{pmatrix} x & y \\ y^* & -x^* \end{pmatrix} \tag{22}$$

with $|x|^2 + |y|^2 = 1$. Substituting this in Eq. (21), we find that $\hat{u}_1^{1\dagger}\hat{u}_1^2 = \hat{\sigma}_x \hat{R}_z(2\phi_1)$ where the rotation operator $\hat{R}_z(\gamma)$ is given as $\hat{R}_z(\gamma) = e^{-i\gamma\hat{\sigma}_z/2}$. Note also that $\hat{u}_1^{1\dagger}\hat{u}_1^2$ is a normal operator. Similar expression can be easily obtained for the i-th player just considering two output states differing in the operators of the i-th player. Next, we take two output states differing in the operators of two players, e.g.,$|\Phi_1\rangle$ and $|\Phi_3\rangle$. Inner product of these states is

$$\langle \Phi_1 | \Phi_3 \rangle = \frac{1}{N} \left(\langle 01 | + \langle 10 | \right) \left(\hat{u}_1^{1\dagger} \hat{u}_1^2 \right) \otimes \left(\hat{u}_2^{1\dagger} \hat{u}_2^2 \right) \left(| 01 \rangle + | 10 \rangle \right)$$

$$= \frac{1}{N} \left(e^{i(\phi_1 - \phi_2)} + e^{-i(\phi_1 - \phi_2)} \right) = \frac{2}{N} \cos(\phi_1 - \phi_2). \qquad (23)$$

The condition of orthogonality requires $\cos(\phi_1 - \phi_2) = 0$, thus $\phi_1 - \phi_2 = n\pi + \pi/2$ where n is an integer. Thus, we have $\phi_j - \phi_k = n\pi + \pi/2$ for any two output states different only in the strategies of j- and k-th players. Then for any three player, we have the set of equations $\chi_{jkm} = \{\phi_j - \phi_k = n\pi + \pi/2, \phi_m - \phi_j = n'\pi + \pi/2, \phi_k - \phi_m = n''\pi + \pi/2\}$ with n, n' and n'' being integers. One can easily notice that the sum of the three equations in this set leads to a contradiction: The sum of the three equations in χ_{jkm} results in $3\pi/2 + m'\pi = 0$ which is impossible for integer $m' = n + n' + n''$. Thus, we conclude that there exists no two operators corresponding to the classical strategies of the players when $| W_N \rangle$ is shared in a two-strategy N-player game. Therefore, the classical game cannot be reproduced and hence the W state cannot be used.

General form of states satisfying the strong reproducibility criterion - Now suppose that for an N-qubit entangled state $| \Phi \rangle$ we have found two operators $\{\hat{u}_i^1, \hat{u}_i^2\}$ implying that we have a situation where the strong reproducibility criterion is satisfied. Let us again take the output states differing in the operators of only one player say $|\Phi_1\rangle = \left(\hat{u}_1^1 \otimes \hat{u}_2^1 \otimes ... \otimes \hat{u}_N^1 \right) | \Phi \rangle$ and $|\Phi_2\rangle = \left(\hat{u}_1^2 \otimes \hat{u}_2^1 \otimes ... \otimes \hat{u}_N^1 \right) | \Phi \rangle$. Taking the inner product and imposing the reproducibility criterion, we find $\langle \Phi | \hat{u}_1^{1\dagger} \hat{u}_1^2 \otimes \hat{I} \otimes ... \otimes \hat{I} | \Phi \rangle = 0$. Since $\hat{u}_1^{1\dagger} \hat{u}_1^2$ is a normal operator, it can be diagonalized by a unitary operator, say $\hat{z}_1$. Thus, we can write $\hat{D}_1 = \hat{z}_1 \hat{u}_1^{1\dagger} \hat{u}_1^2 \hat{z}_1^\dagger = \hat{R}_z(-2\phi_1)$ where $e^{\mp i\phi_1}$ are the eigenvalues of $\hat{u}_1^{1\dagger} \hat{u}_1^2$, and consequently we have

$$\langle \Phi | \hat{z}_1^\dagger \hat{z}_1 \hat{u}_1^{1\dagger} \hat{u}_1^2 \hat{z}_1^\dagger \hat{z}_1 \otimes \hat{I} \otimes \cdots \otimes \hat{I} | \Phi \rangle = \langle \Phi' | \hat{D}_1 \otimes \hat{I} \otimes \cdots \otimes \hat{I} | \Phi' \rangle$$

$$= \langle \Phi' | \hat{R}_z(-2\phi_1) \otimes \hat{I} \otimes \cdots \otimes \hat{I} | \Phi' \rangle = 0, \qquad (24)$$

where $| \Phi' \rangle = \hat{z}_1 \otimes \hat{I} \otimes \cdots \otimes \hat{I} | \Phi \rangle$. Now writing the state $| \Phi' \rangle$ in computational basis as $| \Phi' \rangle = \Sigma_{i_j \in \{0,1\}} c_{i_1 i_2 ... i_N} | i_1 \rangle | i_2 \rangle \cdots | i_N \rangle$, and substituting into Eq. (24), we find

$$\cos \phi_1 + i \left(2\Sigma_{i_j \in \{0,1\}} |c_{0 i_2 ... i_N}|^2 - 1 \right) \sin \phi_1 = 0 \qquad (25)$$

which is satisfied for $\cos \phi_1 = 0$ and $\Sigma_{i_j \in \{0,1\}} |c_{0 i_2 ... i_N}|^2 = 1/2$. From $\cos \phi_1 = 0$, we see that $\phi_1 = n\pi + pi/2$ implying that $\hat{D}_1 = i\sigma_z$. This result can be extended for the other players, therefore we have $\hat{D}_k = i\sigma_z$.

The above equations imply that if the N-qubit state $|\Phi\rangle$ and the operators $\{\hat{u}_k^1, \hat{u}_k^2\}$ satisfy reproducibility criterion, then the state $|\Phi'\rangle = \hat{z}_1 \otimes \hat{z}_2 \otimes \cdots \otimes \hat{z}_N |\Phi\rangle$ and the unitary operators $\{\hat{D}, \hat{I}\}$ should satisfy, too. Neglecting the irrelevant global phase of i of $\hat{D}$, and substituting it into all possible output states written as in Eq. (24), we end up with $2^N - 1$ equalities to be satisfied:

$$\langle\Phi'|\hat{\sigma}_z \otimes \hat{I} \otimes \hat{I} \otimes \cdots \otimes \hat{I}|\Phi'\rangle = 0,$$

$$\langle\Phi'|\hat{I} \otimes \hat{\sigma}_z \otimes \hat{I} \otimes \cdots \otimes \hat{I}|\Phi'\rangle = 0,$$

$$\vdots$$

$$\langle\Psi'|\hat{\sigma}_z \otimes \hat{\sigma}_z \otimes \cdots \otimes \hat{\sigma}_z|\Psi'\rangle = 0. \tag{26}$$

Rewriting these equations in the matrix form yields

$$
\begin{bmatrix}
1 \ldots 1 & -1 \ldots -1 \\
\cdots & \cdots & \cdots \\
& \vdots & \\
& \vdots & \\
1 \ldots 1 & 1 \ldots 1
\end{bmatrix}
\begin{bmatrix}
|c_{00\ldots0}|^2 \\
|c_{00\ldots1}|^2 \\
\vdots \\
\vdots \\
|c_{11\ldots1}|^2
\end{bmatrix}
=
\begin{bmatrix}
0 \\
0 \\
\vdots \\
\vdots \\
1
\end{bmatrix}, \tag{27}
$$

where the last row is the normalization condition. The row vector corresponds to the diagonal elements of $\hat{\sigma}_z^{\{0,1\}} \otimes \cdots \otimes \hat{\sigma}_z^{\{0,1\}}$ where $\hat{\sigma}_z^0$ is defined as $\hat{I}$. Consider the operators $\hat{x}, \hat{y} \in (\hat{\sigma}_z^{\{0,1\}})^{\otimes N}$ where $\hat{x}\hat{y} \in (\hat{\sigma}_z^{\{0,1\}})^{\otimes N}$. Since $\text{Tr}[\hat{\sigma}_z] = 0$, for $\hat{x} \neq \hat{y}$, we have $\text{Tr}[\hat{x}\,\hat{y}] = \text{Tr}[\hat{x}]\,\text{Tr}[\hat{y}] = 0$. Thus any two row vectors are orthogonal to each other, thus the matrix in Eq. (27) has an inverse, and $|c_{i_1 i_2 \ldots i_N}|^2$ are uniquely determined as $1/N$. This implies that if a state satisfies strong reproducibility criterion, then it should be transformed by local unitary operators into the state which contains all possible terms with the same magnitude but different relative phases:

$$|\Phi'\rangle = \Sigma_{i_j \in \{0,1\}} N^{-1/2} e^{i\phi_{i_1 i_2 \ldots i_N}} |i_1\rangle|i_2\rangle \cdots |i_N\rangle. \tag{28}$$

One can easily show that the product state can be transformed into the form of Eq. (28) when all the players apply the Hadamard operator, $\hat{H} = (\hat{\sigma}_x + \hat{\sigma}_z)/\sqrt{2}$. The GHZ state, which also satisfies the strong reproducibility criterion, can be transformed into the same form by $(e^{i\frac{\pi}{4}}\hat{I} + e^{-i\frac{\pi}{4}}\hat{\sigma}_z + \hat{\sigma}_y)/\sqrt{2}$ for one player and $\hat{H}$ for the others. Thus, Bell states, product states and $|GHZ_N\rangle$ states can be used in quantum versions of classical two-strategy multiplayer games. In the same way, one can easily show that $|W_N\rangle$ and Dicke states $|N - m, m\rangle$ except $|2,2\rangle$ and $|1,1\rangle$ cannot be written in the form Eq. (28) and hence cannot be used. The proof for Dicke states can also be obtained in the way we have done in the previous subsection for W-states. We find that the Dicke state $|2, 2\rangle$ can be used in a quantum game by assigning the classical strategies to the operators $\hat{u}^1_{k=1,2,3,4} = \hat{I}$, $\hat{u}^2_{k=1,2,3} = i(\sqrt{2}\hat{\sigma}_z + \hat{\sigma}_x)/\sqrt{3}$, and $\hat{u}^2_4 = i\hat{\sigma}_y$. The eigenvalues of $\hat{u}^{1\dagger}_k\hat{u}^2_k$ are i and $-i$ and they are already in the diagonalized form. For GHZ state, the operators are $\hat{u}^1_k = \hat{I}$ and $\hat{u}^2_k = i\hat{\sigma}_y$ which can also be written in the form $\hat{D}$.

Weak reproducibility criterion and W states - In this section we check whether the above results for strong reproducibility criterion are valid or not for the games where payoffs are the same for some of the possible output states after the players perform their operators. We will consider only the W-states here but a detailed analysis for Dicke states can be found in.[41] The results of such a study requires a classification of multiplayer games because the results would be very much dependent on the structure of payoff matrices. The following observations for the W-states from the above analysis make our task easier:

(i) If the mutual orthogonality condition of the subsets (each subset includes the all output states with the same payoffs) leads to the operator form $\hat{u}^{1\dagger}_k\hat{u}^2_k = \hat{\sigma}_x\hat{R}_z(2\phi_k)$ for any two output states differing only in the strategy of k-th player, then there will be *contradiction* if we obtain the set χ_{jkm} for any three-player-combination (j, k, m). Presence of at least one such set is enough to conclude that there is *contradiction*.

(ii) If the mutual orthogonality condition of the subsets leads to the operator form as in $\hat{u}^{1\dagger}_k\hat{u}^2_k = \hat{\sigma}_x\hat{R}_z(2\phi_k)$ for one and only one player, then there will be *no contradiction* because there will be at least one missing equation in χ_{jkm} for all possible three-player-combinations (j, k, m). Note that such a situation occurs iff 2^N possible outputs are divided into two subsets with equal number of elements. Then the only equations we will obtain are $\phi_1 - \phi_j = n\pi + \pi/2$ for all $j = 2, \cdots, N$.

(iii) If the number of elements in any of the subsets in a game is an odd number, then there will always be *contradiction*

(iv) If there is a set with only two elements which are the outcomes when all the players choose the same strategy, there will be *contradiction*.

If we encounter a *contradiction* in the analysis of a multiplayer game played with W-states, we conclude that W-states cannot be used in the quantum version of that game. Checking these observation on various games, we found out that W-states cannot be used in (1) multiparty extensions of originally two-player two-strategy games such as SH, CG, BoS, BB,

MP, and AG due to contradiction explained in (iii), (2) minority and majority games due to contradiction explained in (i), (3) coordination games due to (i-iv), (4) zero sum gates due to (iv), and (5) symmetric games among which we can list PD, MD, RC, CG, SH, and AG up to six players due to (iii).

As an example, let us analyze the minority game with four-players. In this game, each player chooses between two strategies. The choices are then compared and the player(s) who have made the minority decision are rewarded by one point. If there is an even split, or if all players have made the same choice, then there is no reward. The players with the majority decision loses one point. The structure of this game reflects many common social dilemmas, for example choosing a route in rush hour, choosing which evening to visit an over-crowded bar, or trading in a financial market. Since we are dealing with a four-party case, there are 16 possible outcomes, $\{|\Phi_{\mathcal{L}=1,...,16}\rangle\}$ where the subscript is $\mathcal{L} = 1 + \sum_{i=1}^{4} \ell_i 2^{4-i}$ with $\ell_i = 0$ (or $\ell_i = 1$) when the k-th player chooses the first operator $\hat{u}_k^1$ (or the second operator $\hat{u}_k^2$). For example, for the joint action $\hat{u}_1^1 \hat{u}_2^2 \hat{u}_3^2 \hat{u}_4^1$ where the first and fourth players choose the first, and the second and third players choose the second operators, we find $\ell_{1,4} = 0$ and $\ell_{2,3} = 1$ and consequently $\mathcal{L} = 7$ implying that $|\Phi_7\rangle$ is the output state received by the referee. In the same way, if the output state is $|\Phi_2\rangle$, then we know that $\mathcal{L} = 2$ which is calculated using $\ell_{1,2,3} = 0$ (first, second and third players have chosen the first operator), and $\ell_4 = 1$ (the fourth player has chosen the second operator).

For eight output states out of sixteen possible outputs for the four-player minority game, there is an even-split (half of the players prefers first operator and the other half prefers the second operator) and hence all the players are rewarded with payoff zero. We group these eight states into the set $S_1 = \{|\Phi_{1,4,6,7,10,11,13,16}\rangle\}$. Each player is minority in two cases, therefore the rest of the eight output states can be grouped into $S_2 = \{|\Phi_{2,15}\rangle\}$, $S_3 = \{|\Phi_{3,14}\rangle\}$, $S_4 = \{|\Phi_{5,12}\rangle\}$, and $S_5 = \{|\Phi_{8,9}\rangle\}$ where fourth, third, second and first players are minority players, respectively. Note also that in each of the sets we have output states differing in the operators of all the players. From the set-pair (S_1, S_3) using $\langle\Phi_3|\Phi_1\rangle$, from the set-pair (S_1, S_4) using $\langle\Phi_5|\Phi_1\rangle$, from the set-pair (S_1, S_5) using $\langle\Phi_9|\Phi_1\rangle$, and from the set-pair (S_1, S_2) using $\langle\Phi_2|\Phi_1\rangle$, we find $\hat{u}_k^{1\dagger}\hat{u}_k^2 = \hat{\sigma}_x \hat{R}_z(2\phi_k)$ for $\forall k$. Substituting $\hat{u}_k^{1\dagger}\hat{u}_k^2$ into the expressions obtained from the orthogonality relations of the sets (S_2, S_5), (S_3, S_5) and (S_4, S_5), we see that the set of expressions represented by χ_{234} in (i) is obtained. Therefore, similar contradiction is observed.

For a zero-sum game with N players it is easy to see that a contradiction similar to the one in (iv) occurs. In a zero-sum game with a competitive advantage ξ, there is no winner or loser so all the players receive zero payoffs. Otherwise, t players choosing the first strategy gets ξ/t, and the rest $N - t$ players receive $\xi/(N - t)$. Then the output states can be grouped into $2^N - 1$ different sets. One of these sets will contain the two output states when all players choose the same strategy, and each of the rest will contain only one state. Such an output state partition is exactly what leads to the contradiction in (iv). Therefore, W-states cannot be used in zero-sum games.

Our studies have revealed that the results obtained under the strong reproducibility condition for the W, GHZ, Dicke and Bell states are valid for a large class of classical games whose classical outcomes correspond to the same payoff. We restricted our results to the specific entangled states which have already been created in experiments. The *reproducibility criterion* provides a fair basis to compare quantum versions of games with their classical counterparts because only the states and games which satisfy the criterion in a physical scheme can be compared and played. The analysis shows that a large class of multipartite entangled states cannot be used in the quantum version of classical games; and that the operators that might be used should have a special diagonalized form.

8. Conclusion

In this lecture note, we have attempted to clarify some of the dynamics in the games when the classical games are transferred to quantum domain. The widely used Eisert's scheme is discussed and applied to specific games to show the effects of shared correlations and the set of operators available to the players on the game dynamics. One interesting finding is that for some games entanglement is not needed to arrive at an NE and resolve the dilemma in the game; however it is necessary to find the solution with the highest payoff distribution to the players. Classical correlations in a game may lead to NE and in some cases resolve the dilemma however they never lead to payoffs as high as those in the case of entanglement. The comparison of various games also revealed that quantum advantage does not necessarily take place in all the games; it is strongly dependent on the game payoff matrix. In an attempt to form a fair basis to evaluate the entanglement in games and to compare the game dynamics, we suggested to impose a reproducibility condition which states that when a quantum mechanical setting is used to play the game the classical game should be reproduced in that specific setting. This condition lead us to give general forms of the entangled states and quantum operators which can be used in quantum versions of classical games.

In the rapidly growing quantum information science, it is inescapable to deal with conflict and competitive situations. The expectation is that quantum game theory may help when dealing with such situations. It is also expected that development and understanding of quantum game theory will pave the way to develop efficient quantum algorithms and protocols. It is still not clear how we can utilize the present knowledge learned from the study of game theory in the past years to meet the above expectations. Therefore, there is still much work to be done to fully understand the mechanisms behind the interesting observations in quantum games, and to develop a more fundamental game theory which is associated with quantum mechanics.

Acknowledgements

The authors thank Prof. M. Koashi, Dr. T. Yamamoto and Dr. F. Morikoshi for fruitful discussions, advices and support. Ş. K. Özdemir thanks Prof. M. Nakahara, Dr. R. Rahimi and Dr. T. Aoki for their invitation and kind hospitality during the Summer School on Mathematical Aspects of Quantum Computing 2007 held at Kinki University, Higashi-Osaka, Japan.

References

1. E. Rasmusen, *Games and Information: An Introduction to Game Theory* (Blackwell Pub, Oxford, 2001); R. B. Myerson, *Game Theory: Analysis of Conflict* (Harvard Univ. Press, Cambridge MA, 1997).
2. Neumann John von, *Mathematische Annalen* **100** (1928) 295.
3. Neumann John von and Oscar Morgenstern, *The Theory of Games and Economic Behavior*, (New York: Wiley, 1944).
4. J. Orlin Grabbe, quant-ph/0506219
5. P Moulin, and A. Ivanovic, *Proc. of Int. Conf. on Image Process.* **3** (2001) 975.
6. A. S. Cohen, A. Lapidoth, *IEEE Trans. on Inf. Theory* **48** (2002) 1639.
7. J. Conway, N. Sloane, *IEEE Trans. on Inf. Theory* **32** (1986) 337.
8. X. M. Shen, L. Deng, *IEEE Trans. on Signal Process.* **45** (1997) 1092.
9. J. M. Ettinger, in *2nd Inf. Hiding Workshop, Portland, USA, Apr. 15-17* (1998).
10. S. Pateux, G. Le Guelvouit, *Signal Process.: Image Comm.* **18** (2003) 283.
11. L. A. DaSilva, and V. Srivastava, in *The First Workshop on Games and Emergent Behaviors in Distributed Computing Environments, Birmingham, UK, September* (2004)

12. N. F. Johnson *et. al. Fluct. and Noise Lett.* **2** (2002) L305.
13. J.M.R. Parrondo *et. al., Phy. Rev. Lett.* **85** (2000) 5226.
14. C. F. Lee, N. F. Johnson, *Phys. World* **15(10)** (2002) 259.
15. D. A. Meyer, *Phys. Rev. Lett.* **82** (1999) 1052.
16. J. Eisert, M. Wilkens and M. Lewenstein, *Phys. Rev. Lett.* **83** (1999) 3077; *ibid* **87** (2001) 069802.
17. C. F. Lee, and N. F. Johnson, *Phys. Lett. A* **319** (2003) 429.
18. C. F. Lee, and N. F. Johnson, quant-ph/0203049.
19. C. F. Lee, and N. F. Johnson, *Phys. Rev. A* **67** (2003) 022311.
20. J. Shimamura S. K. Ozdemir, F. Morikoshi and N. Imoto, *Int. J. of Quant. Inf.* **2/1** (2004) 79.
21. M. A. Nielsen and I. L. Chuang, *Quantum Computation and Quantum Information*, (Cambridge university Press, 2000).
22. W. Dür, G. Vidal, and J. I. Cirac, *Phys. Rev. A* **62** (2000) 062314.
23. M. Koashi, V. Bužek, and N. Imoto, *Phys. Rev. A* **62** (2000) 050302(R).
24. H. J. Briegel and R. Raussendorf, *Phys. Rev. Lett.* **86** (2001) 910.
25. R. H. Dicke, *Phys. Rev.* **93** (1954) 99.
26. J. K. Stockton et. al., *Phys. Rev. A* **67** (2003) 022112.
27. S. C. Benjamin and P. M. Hayden, *Phys. Rev. Lett.* **87** (2001) 069801.
28. S. C. Benjamin, P. M. Hayden, *Phys. Rev. A* **64(3)** (2001) 030301(R).
29. L. Marinatto and T. Weber, *Phys. Lett. A* **272** (2000) 291.
30. A. Nawaaz, and A. H. Toor, quant-ph/0409046.
31. T. Cheon, and I. Tsutsui, *Phys. Lett. A* **348** (2006) 147.
32. T. Ichikawa, I. Tsutsui, and T. Cheon, quant-ph/0702167.
33. K. Y. Chen, T. Hogg, and R. Beausoleil, *Quant. Inf. Proc* **1** (2002) 449.
34. A. Iqbal and A. H. Toor, *Phys. Lett. A* **300** (2002) 537.
35. R. Kay, N. F. Johnson, and S. C. Benjamin, quant-ph/0102008.
36. A. Iqbal and A. H. Toor, *Phys. Lett. A* **280** (2001) 249.
37. A. Iqbal and A. H. Toor, *Phys. Lett. A* **294** (2002) 261.
38. A. Iqbal and A. H. Toor, quant-ph/0503176
39. S. K. Ozdemir, J. Shimamura, F. Morikoshi, and N. Imoto, *Phys. Lett. A* **313** (2004) 218.
40. J. Shimamura, *Playing games in quantum realm*, PhD Dissertation, Osaka University, Osaka, Japan, March (2005).
41. S. K. Ozdemir, J. Shimamura, and N. Imoto, *New J. Phys.* **9** (2007) 43.
42. E. W. Piotrowski, and J. Sladkowski, *Physica A* **312** (2002) 208.
43. E. W. Piotrowski, and J. Sladkowski, *Physica A* bf 318 (2003) 505.
44. E. W. Piotrowski, and J. Sladkowski, *Quant. Finance* **4** (2004) 1.
45. G. P. Harmer and D. Abbott, *Nature* **402** (1999) 864.
46. A. Allison and D. Abbott, *Fluct. Noise. Lett.* **2** (2002) L327.
47. A. P. Flitney, J. Ng, and D. Abbott, *Physica A* **314** (2002) 35.
48. A. P. Flitney, and D. Abbott, *Physica A* **324** (2003) 152.
49. Du J, Li H, Xu X, Shi M, Wu J and Han R, *Phys. Rev. Lett.* **88** (2002) 137902.
50. R. Prevedel, A. Stefanov, P. Walther, and A. Zeilinger, *New J. Phys.* **9** (2007) 205.

51. M. Paternostro, M. S. Tame, M. S. Kim, *New J. Phys.* **7** (2005) 226.

52. A. P. Flitney, *Aspects of quantum game theory*, PhD Dissertation, University of Adelaide, Australia, January (2005). *Available at http://ariic.library.unsw.edu.au/adelaide/adt-SUA20051027-072026/.*

53. A. Iqbal, *Studies in the theory of quantum games*, PhD Dissertation, Quaid-i-azam University, Islamabad, Pakistan, September (2004). Available at arXiv: quant-ph/0503176.

54. A. F. Huertas-Rosero, *Classification of quantum symmetric nonzero sum* 2×2 *games in the Eisert scheme*, MSc Thesis, Universidad de Los Andes, Bogota, Colombia, February (2004). Available at arXiv: quant-ph/0402117.

55. J. M. Buchanan, "The Samaritan's Dilemma", in *Altruism, Morality, and Economic Theory*, edited by Edmund S. Phelps, (Russell Sage, New York, 1975).

56. S. J. van Enk and R. Pike, *Phys. Rev. A* **66** (2002) 024306.

57. G. Brassard, quant-ph/0101005.

58. N. F. Johnson, *Phys. Rev. A* **63** (2001) 020302.

59. S. K. Ozdemir, J. Shimamura, F. Morikoshi, and N. Imoto, *Phys. Lett. A* **325** (2004) 104.

60. A. P. Flitney and D. Abbott, *J. Phys. A* **38** (2005) 449.

61. A. P. Flitney and L. C. L. Hollenberg, quant-ph/0510108.

62. Y-J. Han, Y-S. Zhang, and G. C. Guo, *Phys. Lett. A* **295** (2002) 61.

63. J. Shimamura, S. K. Ozdemir, F. Morikoshi and N. Imoto, *Phys. Lett. A* **328** (2004) 20.

64. J. Shimamura, S. K. Ozdemir, F. Morikoshi and N. Imoto, quant-ph/0508105.

QUANTUM ERROR-CORRECTING CODES

MANABU HAGIWARA

National Institute of Advanced Industrial Science and Technology,
Akihabara-Daibiru 11F, 1-18-13 Sotokanda,
Chiyoda-ku, Tokyo, 101-0021 Japan
*E-mail: hagiwara.hagiwara@aist.go.jp

In this review, we discuss quantum error-correcting codes by emphasizing the generalization of classical error-correcting codes. Starting with the explanations of general notions of classical error-correcting codes, we finish the discussion with the construction of quantum codes which are constructed by a pair of practical and trendy classical codes.

Keywords: Quantum Error-Correcting Codes, CSS Codes, Stabilizer Codes, LDPC Codes, Quasi-Cyclic LDPC Codes, Group Theory, Algebraic Combinatorics.

1. Introduction

Quantum error-correcting codes provide detection or correction of the errors which occur in a communication through a noisy quantum channel. Hence the codes protect integrity of a message sent from a sender to a receiver.

In this review, we discuss quantum error-correcting codes in view of generalization of classical error-correcting codes. We give examples of basic theory of classical error-correcting codes and then introduce related quantum counterparts. One of the research trends is to generalize algebraic geometrical codes, e.g. BCH codes, Reed-Solomon codes, and others,[1] to quantum codes in view of mathematical aspects of quantum error-correction codes. In this paper, we focus not on algebraic geometrical codes but on LDPC codes.[2] In particular, the main target of a class of LDPC codes which are generalized to quantum codes is called a class of CSS codes.[3] For construction, the nearly random parity-check matrix is required to classical LDPC codes. On the other hand, a constructive but interesting condition, called twisted condition, is required to CSS codes, which is one of the most popular classes of quantum codes. Thus LDPC codes does not seem to be suitable for generalizations to quantum error-correcting codes. Recently, the notion of quasi-cyclic LDPC codes has been proposed, which changed the reseach trend of quantum error-correcting theory. One of the advantages of quasi-cyclic LDPC code is theoretical one that there are various mathematical tools, e.g. discrete mathematics, finite rings, and others, to describe the properties of LDPC codes. Another advantage of a quasi-cyclic LDPC code is practical one that it has high error-correcting performance with short length. In this review we aim to study the construction of quantum error-correcting code in the framework of generalization of classical quasi-cyclic LDPC codes.

Throughout this paper, we assume that a classical error-correcting code is a binary code and a quantum error-correcting code consists of sequences of two level states.

2. Classical Error-Correcting Codes

2.1. *Introduction to classical error-correcting codes*

Let us consider the following of examples of classical error-correcting codes: A sender would like to transmit a single bit $x \in \{0, 1\}$ to a receiver. It is natural to assume that the communication channel is noisy. Let us assume that the bit-flip error occurs with small but non-negligible probability over the communication channel. The sender transmits x itself over the noisy channel and the receiver obtains a single bit y. It is impossible to distinguish for the receiver that the received message y is the original message x or the bit-flipped message $x + 1$ because of noise. Here $+$ is a bitwise addition modulo 2.

Let us introduce an error-correcting code to this scenario. Imagine that the sender transmits a word xxx instead of the single bit x to the receiver over the noisy channel. This implies that the original word is an element of $C = \{000, 111\}$. If one of the three bits is flipped, then the received word is not an element of the set C. Thus the receiver detects the bit-flip error. Furthermore, the receiver corrects the error by a majority decision. The 3-repetition code is useful for classical bit-flip error-correction if the probability of double bit-flip errors in a word over the channel is negligible. The probability that double bit-flip errors occurring in the word depends on a communication channel. If it is not negligible, we should choose a suitable encoding for communication. For example, the 5-repetition code is one of the candidates. The point of encoding is to introduce some form of redundancy to the original information. Study of good encoding is a fundamental and an interesting but an important problem for error-correcting codes. Not only to find a good encoding but also to find a good decoding algorithm is an important problem.

Shannon shows that high error-correcting performance is achieved by a random code, where a random code is a code whose codewords are chosen randomly.[4,5] However, there remain some problems. The random code requires huge length and it is impossible to correct errors over a random code in real-time. Shannon's result is quite interesting and meaningful from theoretical point of view. However we would like to develop a good error correcting code in realistic meaning, easy encoding, easy decoding, and good error-correcting performance with shorter length. The purpose of error-correcting code theory is to develop a code with "easy encoding", "easy decoding method" and "shorter length".

2.2. Linear codes and parity-check matrices

Let $\mathbb{F}_2$ be a prime field with characteristic 2, i.e. $\mathbb{F}_2 = \{0, 1\}$. A subset of $\mathbb{F}_2 \cup \mathbb{F}_2^2 \cup \mathbb{F}_2^3 \cup \ldots$ is called a **code**. An element of $\mathbb{F}_2 \cup \mathbb{F}_2^2 \cup \mathbb{F}_2^3 \cup \ldots$ is called a **bit sequence**. Hence a code (or a code space) means a set which consists of bit sequences and no mathematical structure is required. If a code C is given, an element of a code is called a **codeword** of C. If a codeword (resp. a bit sequence) belongs to $\mathbb{F}_2^n$ for some positive integer n, then the length of the codeword (resp. the bit sequence) is said to be n. For example, the length of $(0, 0, 0, 0, 0)$ is 5.

A code is called a **block code** if the length of any element of the code is independent of a codeword. Thus a block code C is a subset of $\mathbb{F}_2^n$ for a positive integer n. Note that a block code is a finite set. The positive integer n is called the **length** of the code C. Note that we can consider $\mathbb{F}_2^n$ a vector space over $\mathbb{F}_n$. We call a block code $C(\neq \emptyset)$ **linear** if C is a linear subspace of $\mathbb{F}_2^n$, i.e. $c + c' \in C$ for any $c, c' \in C$. Then the code C is a vector space over $\mathbb{F}_2$. We should note that it is not required that a code be a block code to define a linear code. In fact, we can define an infinite vector space over $\mathbb{F}_2$, and an infinite linear error-correcting code. In this section, we focus on finite linear codes among classical linear codes.

The linearity provides us with benefits in studying error-correcting codes. For example, we might apply theory of linear algebra to error-correcting codes. Now let us define the rank, or dimension, of an error-correcting code. If the dimension of a linear code is k as a vector space over $\mathbb{F}_2$, then the **dimension** of a linear code is said to be k. We call a linear code C an $[n, k]$ **code** if $C \subset \mathbb{F}_2^n$ and the rank of C is k. Another example of benefits of linearity is to develop low-cost error-detection. Let us introduce a parity-check matrix of a linear code to explain the realization of the low-cost error-detection. Since C is a vector subspace of $\mathbb{F}_2^n$, there exists a linear mapping H_C from $\mathbb{F}_2^n$ to $\mathbb{F}_2^{n-k}$ such that

$$C = \ker H_C := \{x \in \mathbb{F}_2^n | H_C x^{\mathrm{T}} = 0\}$$

as a binary matrix of size $n \times (n-k)$ and $\mathrm{rank} H_C = n - k$, where x^{T} is the transposed vector of x. We can represent H_C as a binary matrix of size $(n-k) \times n$, which is called a **parity-check** matrix of C. Conversely, a given binary matrix H_C defines a linear code C by putting $C := \{x \in \mathbb{F}_2^n | H_C x^{\mathrm{T}} = 0\}$. The parity-check matrix H_C is not uniquely determined by a given code C. On the other hand, a linear code C is characterized by the parity-check matrix H_C. In other words, an n-bit sequence x is a codeword if and only if $H_C x^{\mathrm{T}} = 0$. Thus we have a characterization for $x \in \mathbb{F}_2^n$ to be a codeword. Let us go back to error-detection theory. For a received word y, we can detect error by calculating $H_C y^{\mathrm{T}}$. The vector $H_C y^{\mathrm{T}} \in \mathbb{F}_2^{n-k}$ is called the **syndrome** of y associated with H_C. If the syndrome is not 0, then we find that there is at least one error in the received message. An algorithm to calculate the syndrome $H_C y^{\mathrm{T}}$ and an algorithm to compare the syndrome $H_C y^{\mathrm{T}}$ with a zero-vector 0 work in practical-time if the length is not so long.

2.3. Minimum distance and $[n, k, d]$ codes

The weight $\mathrm{wt}(x)$ of $x = (x_1, x_2, \ldots, x_n) \in \mathbb{F}_2^n$ is defined by

$$\mathrm{wt}(x) = \#\{i | x_i = 1\}.$$

The weight $\mathrm{wt}(x)$ is called the **Hamming weight** of x. For example, $\mathrm{wt}(1, 0, 0, 0, 1, 1, 0) = 3$. Next we introduce a distance in $\mathbb{F}_2^n$. For $x, y \in \mathbb{F}_2^n$, we define the distance $d(x, y)$ between x and y by

$$d(x, y) := \mathrm{wt}(x - y).$$

The distance is called the **Hamming distance**. For example, the Hamming distance between $(1, 1, 0, 0, 0)$ and $(1, 0, 0, 0, 1)$ is 2, since $(1, 1, 0, 0, 0) - (1, 0, 0, 0, 1) = (0, 1, 0, 0, 1)$. Note that the triangle inequality $d(x, z) \leq d(x, y) + d(y, z)$ holds for any $x, y, z \in \mathbb{F}_2^n$.

Let S be a subset of $\mathbb{F}_2^n$. For S, the **minimum distance** $d(S)$ of S is defined by

$$d(S) := \min\{d(c, c') | c, c' \in S, c \neq c'\}.$$

For example, if we put $S = \{(1, 1, 0, 0, 0), (0, 1, 1, 0, 1), (0, 0, 1, 1, 0)\}$, the minimum distance $d(S)$ is 3.

Proposition 2.1. *Let C be a linear code. Then*

$$d(C) = \min\{\text{wt}(c)|c \in C, c \neq 0\}.$$

Proof. Let c, c' be codewords of C such that $d(C) = d(c, c')$. By the linearity of C, $c - c'$ is a codeword of C. Therefore $d(C) = d(c, c') = \text{wt}(c - c') \geq \min\{\text{wt}(c)|c \in C, c \neq 0\}$.

Let x be a codeword of C such that $\text{wt}(x) = \min\{\text{wt}(c)|c \in C, c \neq 0\}$. Then $d(C) = \min\{d(c, c')|c, c' \in C, c \neq c'\} \leq d(x, 0) = \text{wt}(x) = \min\{\text{wt}(c)|c \in C, c \neq 0\}$.

Consequently, $d(C) = \min\{\text{wt}(c)|c \in C, c \neq 0\}$ holds. $\square$

A linear code C is called an $[n, k, d]$ **code** if C is a block linear code with the length n, dimension k, and minimum distance d. For example, the 3-repetition code (appearing in Sec. 2.1) is a $[3, 1, 3]$ code.

Example 2.1. Let us define a parity-check matrix H_C by

$$H_C = \begin{pmatrix} 0 & 0 & 0 & 1 & 1 & 1 & 1 \\ 0 & 1 & 1 & 0 & 0 & 1 & 1 \\ 1 & 0 & 1 & 0 & 1 & 0 & 1 \end{pmatrix}.$$

Note that the i th column of H_C is equal to the binary expression of i. The code defined by H_C is called a **Hamming code**.

We show that the Hamming code is a $[7, 4, 3]$ code. Clearly, the length is 7. Because the rows of H_C are linearly independent, the dimension of the code is 4. Thus the Hamming code is a $[7, 4]$ code. We show that the minimum distance of the Hamming code is 3. In other words, there is no codeword c of the Hamming code such that $1 \leq \text{wt}(c) \leq 2$ (Proposition 2.1). Assume that there is a codeword c such that $\text{wt}(c) = 1$. Then $c = e_i$, where e_i denotes a vector such that the i th entry is 1 and all other entries are zero. The syndrome $H_C e_i^T$ is the binary expression of i, in particular, the syndrome is a non-zero vector, which is a contradiction. Assume that there is a codeword c such that $\text{wt}(c) = 2$. Then c is a sum of two elementary vectors e_i and e_j for some i, j. $H_C e_i^T$ and $H_C e_j^T$ are linearly independent for any $1 \leq i < j \leq 7$. Thus $H_C(e_i + e_j)^T$ is a non-zero vector. On the other hand, there is a codeword c such that $\text{wt}(c) = 3$. In fact, $H_C(e_1 + e_2 + e_3)^T = 0$. Therefore, the minimum distance of the Hamming code is 3.

Remark 2.1. In general, it is difficult to determine the minimum distance of a given code. In fact, determination of the minimum distance for a given linear code is known as a NP-hard problem.

2.4. Encoding

Proposition 2.2. *Let C be a linear code. Then there is a binary matrix G such that $C = \{xG | x \in \mathbb{F}_2^k\}$. The matrix G is called a **generator matrix**.*

Proof. Let k be the dimension of C. Then there are k linearly independent vectors $g_1, g_2, \ldots, g_k$. Define a matrix G by

$$G = \begin{pmatrix} g_1 \\ g_2 \\ \vdots \\ g_k \end{pmatrix}.$$

It is easy to verify that $g_1, g_2, \ldots, g_k$ form a basis of C. In other words, $g_1, g_2, \ldots, g_k$ span the code C. Therefore G is a generator matrix. $\square$

By using a generator matrix G of a linear code C, we have a simple encoding which is multiplication of a message x and the generator matrix G.

An $[n, k]$ code is called **systematic** if the first k bits of a codeword coinside with the original message. For a systematic code, it is easy to recover the original message from a codeword after error-correction. In fact, we have only to delete the last $n - k$ bits to recover the original message. The last $n - k$ bits of a codeword of an $[n, k]$ systematic code are called **check symbols**.

If the right-side block of a parity-check matrix is a lower triangular matrix, then the linear code C of the parity-check matrix is systematic. We can encode C as a systematic code by using the parity-check matrix. Now, we show an example of encoding by using the parity-check matrix. Let us permute the columns of the parity-check matrix in Example 2.1 as follows:

$$H' = \begin{pmatrix} 0\,1\,1\,1\,1\,0\,0 \\ 1\,0\,1\,1\,0\,1\,0 \\ 0\,1\,0\,1\,0\,1\,1 \end{pmatrix}.$$

Then the linear code C' associated with H' is a $[7, 4]$ code. Thus, the information bits consist of four bits. Let $x = (x_1, x_2, x_3, x_4)$ be a four-bit sequence. We define a parity-bit x_5 by putting $x_5 = x_2 + x_3 + x_4$. This definition of x_5 comes from the first row of H'. By a similar way, we define recursively $x_6 := x_1 + x_3 + x_4$ and $x_7 := x_2 + x_4 + x_6$. By the construction of x_5, x_6, x_7, $(x_1, x_2, \ldots, x_7)$ is a codeword of C'.

2.5. *Decoding*

In this review, we use the term "decoding method" to denote error-correction algorithm in this article. We should remember that the decoding method in a restricted sense is to recover the information bits from the codeword. It is known that to decode a given linear code is a NP-hard problem. Thus the decoding algorithm does not work in real time if the length of a given code is very long. However, if we assume some structures for a code, then we might find the decoding algorithm which works practically. First we introduce the decoding algorithms for the following two linear codes: the 3-repetition code and the Hamming code (Example 2.1). Next, we introduce the two general decoding algorithms, called the bounded distance decoding algorithm and the maximum likelihood decoding algorithm. Both decoding algorithms are applicable to any linear code. We should remember that it might not work practically for a general linear code.

Let us explain a decoding algorithm for the 3-repetition code which appeared in Sec. 2.1.

- *Decoding algorithm for the 3-repetition code.*
Input Received word y, which consists of three bits.
Output 000 or 111.
Step 1 Count the number of 1 in the input y.
Step 2 If 1 appears two times or three times in y, then the output is 111.
Step 3 If otherwise, then the output is 000.

As was mentioned in Sec. 2.1, the decoding algorithm works successfully against a single bit-flip error.

The standard decoding algorithm of the Hamming code is to determine the error-position from a syndrome.

- *Decoding algorithm for the Hamming code.*
Input Received word y which consists of seven bits.
Output Estimated-error position: null or positive integer i, where $1 \leq i \leq 7$.
Step 1 Calculate the syndrome of y associated with H (See Example 2.1).
Step 2 If the syndrome y is a zero-vector, then output null.
Step 3 If the syndrome is a binary expression of some integer $1 \leq i \leq 7$, then output i.

If the output is null, we regard that the received word y does not contain errors. If the output is a positive integer i, then we regard that the received word y contains a single bit error e_i. We shall explain the relation between the output of the algorithm and the estimated error-vector. Assume that the single bit-flip error e_i occurs on a codeword c. Then the syndrome is:

$H(c + e_i) = Hc + He_i = He_i =$ the binary expression of i.

Thus the above decoding algorithm works as a single bit-flip error correction algorithm. We note that we can correct a single bit-flip error without the received word if the syndrome is given.

Next we introduce the **bounded distance decoding**. Let C be a linear code with its minimum distance $d(C) = d$.

- *Bounded Distance Decoding*

Input Received word y.

Output Estimated codeword or null

Step 1 Choose a codeword $c \in C$.

Step 2 Calculate the Hamming distance $d(c, y)$.

Step 3 If the Hamming distance $d(c, y)$ is less than or equal to $(d-1)/2$, then output c and stop this algorithm. If not, go to Step 4.

Step 4 If all of the codewords in C are tested by Steps 2 and 3 and failed to produce any output in Step 3, then output null and stop this algorithm.

Step 5 Go back to Step 1.

For a linear code C with its minimum distance d, the bounded distance decoding works well as an error-correction algorithm. We explain the reason why the bounded distance decoding works (See Figure 1). For any $c \in C$, we put a set $P_c := \{x \in \mathbb{F}_2^n | \mathrm{wt}(c - x) \leq (d-1)/2\}$. We claim that $P_c \cap P_{c'} = \emptyset$ if $c \neq c'$. If there is a vector $x \in P_c \cap P_{c'}$ for some c, c', then $\mathrm{wt}(c - c') = d(c, c') \leq d(c, x) + d(x, c') \leq (d-1)/2 + (d-1)/2 < d$. This is a contradiction. Therefore, for a codeword $c \in C$ and an error-vector e, if the received word is $y = c + e$ and $\mathrm{wt}(e) \leq (d-1)/2$, then the algorithm outputs c.

Next we introduce the **maximum likelihood decoding**. In general, a communication channel is defined by using the term of probability distribution of noise. Denote the probability that a variable x is equal to d by $\Pr[x = d]$. The posteriori probability $\Pr[y = d' | x = d]$ is defined by

$$\Pr[y = d' | x = d] = \frac{\Pr[x = d, y = d']}{\Pr[x = d]}.$$

For example, a **binary symmetric channel (BSC)** is a channel with a probability distribution of bit-flip error such that the probability distribution satisfies

$$\Pr[y = 0 | x = 0] = \Pr[y = 1 | x = 1],$$

$$\Pr[y = 1 | x = 0] = \Pr[y = 0 | x = 1],$$

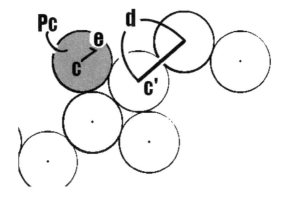

Fig. 1. Schematic idea of bounded distance decoding: Each circle is P_c for some code-word c.

$$\Pr[y = 0|x = 0] + \Pr[y = 1|x = 0] = 1,$$

$$\Pr[y = 0|x = 1] + \Pr[y = 1|x = 1] = 1,$$

and each bit-flip error occurs independently (See Figure 2), where x is the input bit and y is the received bit. The maximum likelihood decoding is to

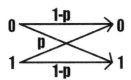

Fig. 2. Image for Binary Symmetric Channel

find the code word c for the received word y such that

$$\Pr[\text{received word} = y|x = c] = \max_{c' \in C} \Pr[\text{received word} = y|x = c'].$$

The maximum likelihood decoding seems to be the best decoding method for error-correction. However, for a given linear code, to calculate the pos-teriori probability for all of codeword of a code is impossible in real time if the code has numbers of codewords.

2.6. *Bit error-rate and block error-rate*

The way of evaluation for error-correcting codes depends on the application. Let us consider a mobile phone, for example. Here we do not require the fidelity of the encoded voice. What is essential for the receiver is to understand the content of the speaker's message. Even if the users are under noisy circumstances, the receiver may understand what the speaker says. On the other hand, the accuracy of each bit is required for a magnetic disk of a personal computer If only one bit in a magnetic disk is broken, then we might never read important data in the disk again.

Two of typical (or widely used) criteria of evaluations for error-correcting codes are values of bit error rate and block error rate. The bit error-rate (BER) is the expectation value of the bit-flip error probability for a bit in the estimated word after a decoding algorithm. In other words,

$$\mathrm{BER} = \mathrm{Ex}_{c=(c_1,c_2,\ldots,c_n)\in C}\left[\frac{\sum_{i=1}^n \Pr[c_i \neq c'_i]}{n}\right],$$

where n is the length of the code and c'_i is the i th bit of the estimated word with the received word c. The **block error-rate** (BLER) is the expectation value of the error probability for a decoding algorithm. In other words,

$$\mathrm{BLER} = \mathrm{Ex}_{c\in C}\left[\Pr[c \neq c']\right],$$

where c is the original encoded message, and c' is the estimated word.

Remark 2.2. It is known that BLER of a classical code is closely related to the fidelity of quantum communication appearing in a quantum code called a CSS code. A CSS code is defined and is investigated later in this review (Sec. 4).

Example 2.2. Let us calculate BLER of the 3-repetition code (see Sec. 2.1) over BSC channel with the cross-over probability p. If no bit-flip error occurs in a transmission, then a block error does not occur. If a single bit-flip error occurs in a transmission, then a block error does not occur either. On the other hand, if two or more error occur in a transmission, then a block error occurs. Thus BLER is:

$$1 - (1 - p)^3 - 3p(1 - p)^2 = p^3 + 3p^2(1 - p).$$

We leave a calculation of BER of the 3-repetition code to for readers.

3. How to Generalize Classical Error-Correcting Codes to Quantum Codes

In this section, we introduce widely used terminologies to study quantum error-correcting codes. Important classical objects such as parity-check matrix, code space, syndrome, and others, which appeared in the previous section, are generalized to quantum error-correcting codes in a natural way. Throughout this paper, matrices I, X, Z, and Y denote the following:

$$I = \begin{pmatrix} 1 & 0 \\ 0 & 1 \end{pmatrix}, X = \begin{pmatrix} 0 & 1 \\ 1 & 0 \end{pmatrix},$$

$$Z = \begin{pmatrix} 1 & 0 \\ 0 & -1 \end{pmatrix}, \text{ and } Y = iXZ.$$

3.1. Parity-check measurements and stabilizer codes

Let $\mathcal{H}_1, \mathcal{H}_2, \ldots, \mathcal{H}_{n-k}$ be Hermitean operators such that

$$\mathcal{H}_i = \mathcal{H}_{i,1} \otimes \mathcal{H}_{i,2} \otimes \cdots \otimes \mathcal{H}_{i,n}$$

and

$$\mathcal{H}_i \mathcal{H}_{i'} = \mathcal{H}_{i'} \mathcal{H}_i$$

for any $1 \leq i, i' \leq n-k$, where $\mathcal{H}_{i,j} \in \{I, X, Z, Y\}$ and $\otimes$ is the tensor product operation. The set of the Hermitean operators $\mathcal{H}_1, \mathcal{H}_2, \ldots, \mathcal{H}_{n-k}$ are called a **parity-check measurement**. By the condition of $\mathcal{H}_i$, its eigenvalues are 1 and -1. A **stabilizer code**, which is a quantum error-correcting code, is an eigenspace such that the measurement outcomes of all $\mathcal{H}_i$ are 1. Thus a stabilizer code is a subspace of $\mathbb{C}^{2 \otimes n}$. For a quantum state in $\mathbb{C}^{2 \otimes n}$, the outcomes of a parity-check measurement are called the **syndrome**. As the names suggest, "parity-check measurement" (resp. "syndrome") is closely related to "parity-check matrix" (resp. "syndrome") in the previous section. We explain relations between these terms later (Proposition 3.2).

Next proposition shows the dimension of a stabilizer code and its mother set. In classical error-correcting codes, the dimension of an $[n, k]$ code is k and the dimension of its mother set is n. For quantum error-correcting codes, the dimension of the mother set $\mathbb{C}^{2 \otimes n}$ is 2^n and the dimension of a stabilizer code is shown in the following:

Proposition 3.1. *If each of the Hermitean operators $\mathcal{H}_i$ is not included in a group generated by $\mathcal{H}_j (j \neq i)$, then the dimension of an eigenspace with respect to a fixed syndrome is 2^k. In particular, the dimension of the related stabilizer code is 2^k.*

Proof. Denote the dimension of eigenspace V_s associated with a syndrome $s := (s_1, s_2, \ldots, s_{n-k})$ by $\dim(V_s)$, where s_i is the measurement outcome of the operator $\mathcal{H}_i$. For $I \subset \{1, 2, \ldots, n-k\}$, we define a Hermitian operator $\mathcal{H}_I$ by putting $\mathcal{H}_I := \prod_{i \in I} \mathcal{H}_i$. Similarly, we put $s_I := \prod_{i \in I} s_i$.

There are two eigenspaces of $\mathcal{H}_I$ and their eigenvalues are 1 and -1. The eigenspaces of $\mathcal{H}_I$ associated with an eigenvalue e is the linear subspace $V_{I,e}$ spanned by $\{V_s | s_I = e\}$. Thus

$$\dim V_{I,e} = \sum_{s, s_I = e} \dim(V_s).$$

The dimension of $V_{I,1}$ is equal to the dimension of $V_{I,-1}$ for $I \neq \emptyset$, since the trace of $\mathcal{H}_I = \dim V_{I,1} - \dim V_{I,-1} = 0$. It implies:

$$\dim V_{I,1} = \dim V_{I,-1} = 2^{n-1}.$$

Next, let us calculate $\sum_{I \subset \{1,\ldots,n-k\}, I \neq \emptyset} \dim V_{I,s_I}$. Since $\dim V_{I,1} = \dim V_{I,-1} = 2^{n-1}$, we have

$$\sum_{I \subset \{1,\ldots,n-k\}, I \neq \emptyset} \dim V_{I,s_I} = 2^{n-1}(2^{n-k} - 1).$$

On the other hand,

$$\sum_{I \subset \{1,\ldots,n-k\}, I \neq \emptyset} \dim V_{I,s_I} = \sum_{I \subset \{1,\ldots,n-k\}, I \neq \emptyset} \left(\sum_{r \in \{1,-1\}^{n-k}, r_I = s_I} \dim(V_r) \right)$$

$$= \sum_{I \subset \{1,\ldots,n-k\}, I \neq \emptyset} \left(\dim(V_s) + \sum_{r \in \{1,-1\}^{n-k}, r_I = s_I, r \neq s} \dim(V_r) \right)$$

$$= (2^{n-k} - 1) \dim(V_s) + \sum_{r \neq s} (2^{n-k-1} - 1) \dim(V_r)$$

$$= 2^{n-k-1} \dim(V_s) + (2^{n-k-1} - 1) 2^n.$$

Thus $\dim(V_s) = 2^k$. $\qquad\square$

By the following proposition, the readers may agree that a parity-check measurement is regarded as a generalization of a parity-check matrix of a classical code.

Proposition 3.2. *Let* $\mathcal{H}_1 = \mathcal{H}_{1,1} \otimes \mathcal{H}_{1,2} \otimes \cdots \otimes \mathcal{H}_{1,n}, \mathcal{H}_2 = \mathcal{H}_{2,1} \otimes \mathcal{H}_{2,2} \otimes \cdots \otimes \mathcal{H}_{2,n}, \ldots, \mathcal{H}_{n-k} = \mathcal{H}_{n-k,1} \otimes \mathcal{H}_{n-k,2} \otimes \cdots \otimes \mathcal{H}_{n-k,n}$ *be a parity-check measurement such that* $\mathcal{H}_{i,j} \in \{I, Z\}$. *Take a (classical) parity-check matrix* $H = (H_{i,j})$ *by* $H_{i,j} = 1$ *if* $\mathcal{H}_{i,j} = Z$ *and* $H_{i,j} = 0$ *if* $\mathcal{H}_{i,j} = I$. *Then the* *stabilizer code associated with the parity-check measurement consists of a linear combination of quantum states $|c\rangle$, where c is a codeword with a parity-check matrix H.*

Proof. It is obvious that a quantum state $|c\rangle$ is an element of the stabilizer code. Thus a linear subspace spanned by $\{|c\rangle | c \in C\}$ is contained in the stabilizer code. These spaces are identified by comparing the dimension of them. □

3.2. *Encoding*

Encoding should be a unitary operation. Thus the dimension of information qubits must be equal to the dimension of encoded qubits. On the other hand, encoding is a map from a 2^k-dimensional space to $\mathbb{C}^{2 \otimes n}$. Therefore we need to add $n - k$ initial states $|0\rangle$ to the information (i.e. pre-encoded) qubits. We would like to leave the following simple question as an exercise for readers: Generalize a classical encoding by the lower triangle parity-check matrix to an encoding for a quantum error-correction code by a parity-check measurement.

3.3. *Decoding*

By the law of quantum mechanics, it is impossible to measure the received state itself which is transmitted over a noisy channel. Instead of the received state which is a quantum codeword of a stabilizer code, we can measure a syndrome associated with the parity-check measurement. We have already seen an error-correction method by using syndrome in Example 2.1. It expects that we might correct quantum errors by using the syndrome, if the stabilizer code is well designed.

4. CSS Codes

A stabilizer code is called a **CSS** code if there exists a parity-check measurement such that each observable $\mathcal{H}_i$ consists of tensor product of either I, X or one of I, Z.

4.1. *Twisted condition*

By the definition of a CSS code, we can separate the observables to two sets $\{\mathcal{H}_i\}_{i=1...t}$ and $\{\mathcal{H}_i\}_{i=t+1...t+t'}$ for some t, t' such that $\mathcal{H}_i = Z^{hc_{i,1}} \otimes Z^{hc_{i,2}} \otimes \cdots \otimes Z^{hc_{i,n}}$ for $1 \leq i \leq t$ and $\mathcal{H}_{t+i'} = X^{hd_{i',1}} \otimes X^{hd_{i',2}} \otimes \cdots \otimes X^{hd_{i',n}}$ for

$1 \leq i' \leq t'$, where $hc_{i,j}$ and $hd_{i',j} \in \mathbb{F}_2$. Then we can construct two classical parity-check matrices $H_C = \{hc_{i,j}\}_{i=1...t}, H_D = \{hd_{i,j}\}_{i=1...t'}$.

Let C and D be classical linear codes with the parity-check matrices H_C and H_D respectively. By the commutativity of the observables, the classical linear codes C and D satisfy the following condition: each row of H_C and each row of H_D are orthogonal with the inner product $\langle , \rangle$, where $\langle (a_1, a_2, \ldots, a_n), (b_1, b_2, \ldots, b_n) \rangle = \sum a_i b_i$. In fact, observables $\mathcal{H}_i$ and $\mathcal{H}_{t+j}$ are commutative if and only if Z of $\mathcal{H}_i$ and X of $\mathcal{H}_{t+j}$ meet even times. It is equivalent to claim that there are even positions p such that $hc_{i,p} = hd_{j,p} = 1$. In other words,

$$H_C \times H_D^{\mathrm{T}} = 0,$$

where A^{T} is the transposed matrix of a matrix A.

Let us recall that the (classical) dual code of a linear code. For a linear code C, the dual code $C^{\perp}$ is defined by

$$C^{\perp} = \{x \in \mathbb{F}_2^n | \langle x, c \rangle = 0, \forall c \in C\}.$$

In other words, the **dual code** is a linear code generated by rows of the parity-check matrix. Finally, we conclude that the linear codes C and D obtained from parity-check measurements of CSS codes above satisfy the following condition:

$$D^{\perp} \subset C (\Leftrightarrow C^{\perp} \subset D).$$

We call this condition the **twisted condition**.

Proposition 4.1. *Let C (resp. D) be linear codes with a parity-check matrix H_C (resp. H_D). C and D satisfy the twisted condition if and only if H_C and H_D are orthogonal i.e. $H_C \times H_D^{\mathrm{T}} = 0$.*

Proof.

$$D^{\perp} \subset C$$

$$\Longleftrightarrow \forall x \in D^{\perp}, x \in C$$

$$\Longleftrightarrow \forall x \in D^{\perp}, H_C x^{\mathrm{T}} = 0$$

$$\Longleftrightarrow \forall \text{ row } x \text{ of } H_D, H_C x^{\mathrm{T}} = 0$$

$$\Longleftrightarrow H_C \times H_D^{\mathrm{T}} = 0 \qquad \qquad \square$$

4.2. Characterization of code space

Lemma 4.1. *Let D be a classical linear code and $D^\perp$ the dual code of D. For $x \in \mathbb{F}_2^n$, $\sum_{d' \in D^\perp} (-1)^{\langle x, d' \rangle} = \#D^\perp$ (resp. $= 0$), if $x \in D$ (resp. if otherwise), where $\#X$ is the cadinarity of a set X.*

Proof. If $x \in D$, then $\langle x, d' \rangle = 0$ for all $d' \in D^\perp$. Thus

$$\sum_{d' \in D^\perp} (-1)^{\langle x, d' \rangle} = \#D^\perp.$$

If otherwise, there is $d'_0 \in D^\perp$ such that $\langle x, d'_0 \rangle = 1$. Since $\{d_0 + d | d \in D^\perp\} = D^\perp$,

$$\sum_{d' \in D^\perp} (-1)^{\langle x, d' \rangle} = \sum_{d' \in D^\perp} (-1)^{\langle x, d' + d'_0 \rangle} = (-1) \sum_{d' \in D^\perp} (-1)^{\langle x, d' \rangle}.$$

Therefore $\sum_{d' \in D^\perp} (-1)^{\langle x, d' \rangle} = 0$. $\square$

Theorem 4.1. *Let C and D be classical linear codes which satisfy twisted condition. Then the code space of the CSS code, defined by C and D, is spanned by the following quantum states:*

$$\left\{ \sum_{d' \in D^\perp} |c + d'\rangle \right\}_{c \in C}.$$

Proof. Denote the Hadamard operator by H. (Do not confuse the Hadamard operator H and a parity-check matrix H_C.)

Let Q be a spanned by $\{\sum_{d' \in D^\perp} |c + d'\rangle | c \in C\}$. By the definition, the dimension of Q is equal to $\dim C / D^\perp = \dim C - \dim D^\perp$. Remember that the number of observables of the parity-check measurement is $(n - \dim C) + \dim D^\perp$. By Theorem 3.1, the dimension of the stabilizer code is equal to that of Q. Thus it is sufficient to show that a quantum state $|\phi\rangle := \sum_{d' \in D^\perp} |c + d'\rangle$ is a state of the CSS code.

Since $D^\perp \subset C$ and $c + d \in D^\perp$, it is obvious that $\mathcal{H}_i |c + d\rangle = |c + d\rangle$ for the observable $\mathcal{H}_i$ associated with H_C.

Now, we act a Hadamard operator to each qubit of $|\phi\rangle$ as $H^{\otimes n}|\phi\rangle$.

$$
\begin{aligned}
H^{\otimes n}|\phi\rangle &= \sum_{d' \in D^\perp} H^{\otimes n}|c + d'\rangle = 2^{-n/2} \sum_{d' \in D^\perp} \prod (|0\rangle + (-1)^{c_i}|1\rangle) \\
&= 2^{-n/2} \sum_{x \in \mathbb{F}_2^n} \sum_{d' \in D^\perp} (-1)^{\langle x, c+d'\rangle}|x\rangle \\
&= 2^{-n/2} \sum_{x \in \mathbb{F}_2^n} (-1)^{\langle x,c\rangle} \sum_{d' \in D^\perp} (-1)^{\langle x,d'\rangle}|x\rangle \\
&= 2^{-n/2 + \dim D^\perp} \sum_{d \in D} (-1)^{\langle c,d\rangle}|d\rangle.
\end{aligned}
$$

In particular, the last term above is a linear combination of $|d\rangle$ with $d \in D$. Thus for the observable $\mathcal{H}_i$ associated with H_D, we have

$$
\mathcal{H}_i|\phi\rangle = H^{\otimes n}(H^{\otimes n}\mathcal{H}_i H^{\otimes n})(H^{\otimes n}|\phi\rangle) = H^{\otimes n}(H^{\otimes n}|\phi\rangle) = |\phi\rangle.
$$

Hence, Q is included in the CSS code. □

4.3. Correctable error

Let $\mathcal{E}$ and $\mathcal{R}$ be quantum operators. We call $\mathcal{E}$ a set of **correctable errors** associated with $\mathcal{R}$ if there exists $\alpha \in \mathbb{C}$ such that $(\mathcal{R} \circ \mathcal{E})(\rho) = \alpha\rho$ for any codeword state ρ of quantum error-correcting code. The operator $\mathcal{R}$ is called the **recovery operator**. By a recovery operator, it measures a syndrome of ρ or its partial information. And then, it performs quantum operations which depend on the syndrome or the partial information.

Proposition 4.2. Let $\mathbb{E}_X$ be a subset of $\mathbb{F}_2^n$ and $\mathcal{E}_X$ a subset of $\mathbb{F}_2^n$ defined by $\mathcal{E}_X = \{X^{\otimes e} = X^{e_1} \otimes X^{e_2} \otimes \ldots X^{e_n} | e = (e_1, \ldots, e_n) \in \mathbb{E}_X\}$. It is a subset of bit-flip errors associated with $\mathbb{E}_X$.

Then $\mathcal{E}_X$ is a set of correctable bit-flip errors associated with some recovery operator $\mathcal{R}$ if and only if $e - e'$ is not in $C \setminus D^\perp$ for any $e, e' \in \mathbb{E}_X$.

Proof. If $e - e'$ is not an element of C for any $e' \in \mathbb{E}_X$, then the syndrome $H_C e^\mathrm{T}$ and $H_C (e')^\mathrm{T}$ are different from each other. Thus we can define a map r from the syndrome space to $\mathbb{E}_X$ such that $r(H_C e^\mathrm{T}) = e$. Therefore we can construct a recovery operator $\mathcal{R}$ such that $\mathcal{R} \circ X^e(\rho) = \rho$ for any $X^e \in \mathcal{E}_X$.

Conversely, assume that there exists $e' \in \mathcal{E}_X$ such that $e - e' \in C$. Then the syndrome of $e - e'$ associated with H_C is 0 because $e - e' \in C$. This implies $H_C e = H_C e'$. Since the behavior of a recovery operator depends on the syndrome, $(\mathcal{R} \circ X^e)(\rho) = (\mathcal{R} \circ X^{e'})(\rho)$ holds for any encoded state ρ.

If $e - e' \in D^{\perp} \subset C$, then $X^e \rho = X^{e'} \rho$ holds by the structure of a CSS code. Thus $(\mathcal{R} \circ X^e)(\rho) = (\mathcal{R} \circ X^{e'})(\rho)$ holds.

Assume $e - e'$ is not in $D^{\perp}$ but $e - e' \in C$ and assume there is a recovery operator $\mathcal{R}$ such that $(\mathcal{R} \circ X^e)(\rho) = \rho$ and $(\mathcal{R} \circ X^{e'})(\rho) = \rho$ for any encoded state ρ. Put $|\phi\rangle = \sum_{d' \in D^{\perp}} |d'\rangle$ and $|\psi\rangle = \sum_{d' \in D^{\perp}} |d' + e - e'\rangle$. Since $e - e'$ is not in $D^{\perp}$, $|\phi\rangle \neq |\psi\rangle$ hold. On the other hand, $|\phi\rangle = (\mathcal{R} \circ X^e)(|\phi\rangle) = (\mathcal{R} \circ X^{e'})(|\psi\rangle) = |\psi\rangle$ because $X^e |\phi\rangle = X^{e'} |\psi\rangle$ holds. It is a contradiction.

$\square$

By a similar argument, we have the following:

Proposition 4.3. *Let $\mathbb{E}_Z$ be a subset of $\mathbb{F}_2^n$ and $\mathcal{E}_Z$ a subset of $\mathbb{F}_2^n$ defined by $\mathcal{E}_Z = \{Z^{\otimes e} = Z^{e_1} \otimes Z^{e_2} \otimes \ldots Z^{e_n} | e = (e_1, \ldots, e_n) \in \mathbb{E}_Z\}$ is a subset of bit-flip errors associated with $\mathbb{E}_Z$.*

Then $\mathcal{E}_Z$ is a set of correctable phase-flip errors if and only if $e - e'$ is not in $D \setminus C^{\perp}$ for any $e, e' \in \mathbb{E}_Z$.

Because of the independence of syndromes associated with H_C and H_D, we have the following:

Proposition 4.4. *Let Q be a CSS code. Let $\mathbb{E}_X$ and $\mathbb{E}_Z$ be subsets of $\mathbb{F}_2^n$. Define $\mathcal{E}_X := \{X^{\otimes e} | e \in \mathbb{E}_X\}$ and $\mathcal{E}_Z := \{Z^{\otimes e} | e \in \mathbb{E}_Z\}$. If $\mathcal{E}_X$ and $\mathcal{E}_Z$ are sets of correctable errors for Q, then $\mathcal{E} := \{X^{\otimes e} Z^{\otimes e'} | e \in \mathbb{E}_X, e' \in \mathbb{E}_Z\}$ is also a set of correctable errors for Q.*

By the linearity of a recovery operator, we have the following:

Theorem 4.2. *Let $\mathcal{E} = \{E_i\}$ be a set of correctable errors and $\mathcal{R}$ a set of recovery operators for a quantum code Q. Define a set $F := \sum_i m_i E_i$ for $m_i \in \mathbb{C}$. Then F is a correctable errors by the same recovery operator $\mathcal{R}$.*

Remember that any unitary operator over $\mathbb{C}^2$ is a linear combination of I, X, Z and Y. By Theorem 4.2, if we prove that the errors I, X, Z and Y are correctable for a fixed position, then any unitary error for the position is correctable. Furthermore, if any t-bit error is correctable on classical error-correcting codes C and D, then any t-quantum error is correctable on the CSS code associated with C and D.

4.4. 7-Qubit code (quantum Hamming code)

In this subsection, we would like to introduce an example of a CSS code. The example is a generalization of the classical Hamming code.

Proposition 4.5. *Let C and D be Hamming codes (See Example 2.1). Then C and D satisfy the twisted condition.*

Proof. Let H be a parity-check matrix of the Hamming code in Example 2.1. Then the matrix H is a parity-check matrix H_C (resp. H_D) of C (resp. D). It is easy to verify $H_C \times H_D^T = 0$. $\square$

The Hamming code is a $[7, 4, 3]$ code. In particular, it can correct any single bit-flip error. Thus it should be possible to correct any quantum single position error with a CSS code Q derived from a pair of Hamming codes. We call the CSS code a **7-qubit code** (or a **quantum Hamming code**).

In other words, X_i is a tensor product of seven unitary matrices with X in the i th position and the others are identity matrices. By a similar way, we define Z_i and $(XZ)_i$. Let E be a single quantum error in the 7-qubit code on the i th position. Then $E = aI^{\otimes 7} + bX_i + cZ_i + dXZ_i$, for some $a, b, c, d \in \mathbb{C}$. For a quantum message $|\phi\rangle$, the received message with a quantum error E is

$$E|\phi\rangle = a|\phi\rangle + bX_i|\phi\rangle + cZ_i|\phi\rangle + dXZ_i|\phi\rangle.$$

The syndrome s_X of a parity-check measurement associated with H_C is $(1, 1, 1)$ or $((-1)^{i1}, (-1)^{i2}, (-1)^{i3})$, where (i_1, i_2, i_3) is the binary expression of i. Similarly, the syndrome s_Z of a parity-check measurement associated with H_D is also $(1, 1, 1)$ or $((-1)^{i1}, (-1)^{i2}, (-1)^{i3})$. Thus there are four possibility of the syndrome. For each possibility, we can recover the original state $|\phi\rangle$ by the following:

- If $s_X = s_Z = (1, 1, 1)$, then the quantum state after the measurement is $|\phi\rangle$. Thus we recover the original state.
- If $s_X = ((-1)^{i1}, (-1)^{i2}, (-1)^{i3})$ and $s_Z = (1, 1, 1)$, then the quantum state after the measurement is $X_i|\phi\rangle$. Thus we can recover the original state by multiplying X_i to the state.
- If $s_X = (1, 1, 1)$ and $s_Z = ((-1)^{i1}, (-1)^{i2}, (-1)^{i3})$, then the quantum state after the measurement is $Z_i|\phi\rangle$. Thus we can recoever the original state by multiplying Z_i to the state.
- If $s_X = ((-1)^{i1}, (-1)^{i2}, (-1)^{i3})$ and $s_Z = ((-1)^{i1}, (-1)^{i2}, (-1)^{i3})$, then the quantum state after the measurement is $XZ_i|\phi\rangle$. Thus we can recover the original state by multiplying XZ_i to the state.

We would like to leave the following excise for readers: Let Q be a 7-qubit code. Put $Q_{i,0} := X_iQ, Q_{0,j} := Z_jQ$ and $Q_{i,j} := X_iZ_jQ$ for $1 \leq i, j \leq 7$.

Then 64 quantum subspaces $Q, Q_{i,0}, Q_{0,j}, Q_{i,j} (1 \leq i, j \leq 7)$ are orthogonal to each other.

5. Quantum LDPC Codes

5.1. *LDPC codes and sum-product decoding*

LDPC codes, which are introduced by Gallager, are classical linear codes and their broader-sense definition is a code whose parity-check matrix has fewer 1's than 0's.[2] LDPC stands for "Low-Density Parity-Check". LDPC codes achieve outstanding performance. Its performance approaches the maximum likelihood decoding over some channels with practical assumptions.

The standard decoding algorithm for LDPC codes is called **sum-product decoding**. We introduce the sum-product decoding and explain that it is applicable to a CSS code with a sparse parity-check measurement. We do not mention why the combination of LDPC codes and sum-product decoding works well. Interested readers may verify the high-performance of this combination by computer simulations.

We denote the (m, n)-entry of a binary matrix H by $H_{m,n}$, and we put $A(m) := \{n | H_{m,n} = 1\}, B(n) := \{m : H_{m,n} = 1\}$. We assume that we know the error probability distribution of a communication channel, the noise of the channel does not depend on the original message and the noise occurs independently in each bit. By the syndrome of a received message, sum-product (syndrome) decoding outputs an error vector. The algorithm of sum-product (syndrome) decoding is the following:

- Input: Parity-check matrix H, and syndrome $s = (s_1, s_2, \ldots, s_M)^{\mathrm{T}}$ of the received word y i.e. $s = Hy^{\mathrm{T}}$.
- Output: Estimated error vector $(e_1, e_2, \ldots, e_N)$.
- Step 1: For a pair (m, n) with $H_{m,n} = 1$, put $q_{m,n}(0) = 1/2, q_{m,n}(1) = 1/2$, set the iteration counter to $l = 1$ and introduce the maximum number of iteration $l_{\max}$.
- Step 2: For $m = 1, 2, \ldots, M$ and all of the pairs (m, n) with $H_{m,n} = 1$, update $r_{m,n}(0), r_{m,n}(1)$ by the following formulae:

$$r_{m,n}(0) = K \sum_{\substack{c_i \in \{0,1\}, \\ i \in A_{m,n} \setminus n, \\ \sum c_i = s_m}} \prod_{n' \in A(m) \setminus n} q_{m,n'}(c_{n'}) \Pr(0 | x_{n'} = c_{n'})$$

$$r_{m,n}(1) = K \sum_{\substack{c_i \in \{0,1\}, \\ i \in A_{m,n}\setminus n, \\ \sum c_i = 1 - s_m}} \prod_{n' \in A(m)\setminus n} q_{m,n'}(c_{n'}) \Pr(0|x_{n'} = c_{n'}),$$

where K is a constant such that $r_{m,n}(0) + r_{m,n}(1) = 1$.

- Step 3: For $n = 1, 2, \ldots, N$ and all of pairs (m, n) with $H_{m,n} = 1$, update $q_{m,n}(0)$ and $q_{m,n}(1)$ by the following formula: $q_{m,n}(i) = K' \prod_{m' \in B(n)\setminus m} r_{m',n}(i)$, where $i = 0, 1$ and K' is a constant such that $q_{m,n}(0) + q_{m,n}(1) = 1$.

- Step 4: For $n = 1, 2, \ldots, N$, calculate the following:

$$Q_n(i) = K'' \Pr(y_n | x_n = i) \prod_{m' \in B(n)} r_{m',n}(i)$$

$$\hat{e}_n := 0, \text{if } Q_n(0) \geq Q_n(1)$$

$$\hat{e}_n := 1, \text{otherwise},$$

where $i = 0, 1$ and K'' is a constant such that $Q_n(0) + Q_n(1) = 1$. We call $(\hat{e}_1, \hat{e}_2, \ldots, \hat{e}_N)$ a **temporary estimated error vector**.

- Step 5: Check if the temporary estimated error vector satisfies syndrome check. In other words, if

$$(\hat{e}_1, \hat{e}_2, \ldots, \hat{e}_N) H^{\mathrm{T}} = (s_1, s_2, \ldots, s_M)$$

holds then the algorithm outputs $(\hat{c}_1, \hat{c}_2, \ldots, \hat{c}_N)$ as an estimated error vector and the algorithm stops.

- Step 6: If $l < l_{\max}$, then $l := l + 1$ and return to step 2. If $l = l_{\max}$, then the algorithm outputs $(\hat{e}_1, \hat{e}_2, \ldots, \hat{e}_N)$ as an estimated error vector and the algorithm stops.

By the definition of sum-product algorithm, the output depends on a parity-check matrix. The error-correcting performance changes according to the parity-check matrix, even when the code space is the same.

5.2. Application of the sum-product algorithm for the error-correction of CSS codes

Let C and D be LDPC codes which satisfy the twisted condition and let H_C and H_D be their low-density parity-check matrices, respectively. As we have remarked, the sum-product algorithm outputs an error-vector from a syndrome. By the definition of a CSS code, we can decompose the parity-check measurement into two parts, which are the Z-part associated with H_C, and the X-part associated with H_D.

After receiving a quantum message, we measure it by the parity-check measurement and obtain the syndrome. We decompose the syndrome into two parts s_Z and s_X according to H_C and H_D. We input the parity-check matrix H_C and one of the syndrome s_Z to the sum-product algorithm. Then the algorithm outputs the estimated error-vector $eC = (eC_1, eC_2, \ldots, eC_n)$. We act $X^{eC} = X^{eC_1} \otimes X^{eC_2} \otimes X^{eC_n}$ to the received quantum state. This implies we expect to correct the bit-flip error. Next we input H_D and s_Z to the sum-product algorithm and obtain the estimated error-vector eD. Then we act Z^{eD} to the received quantum state. Consequently we expect that the received state is corrected to the original quantum message.

5.3. Classical and quantum quasi-cyclic LDPC codes

Let $I(\infty)$ be a zero matrix and $I(1)$ a circulant permutation matrix, i.e.

$$I(1) = \begin{pmatrix} & 1 & & & \\ & & 1 & & \\ & & & \ddots & \\ & & & & 1 \\ 1 & & & & \end{pmatrix}.$$

Put $I(b) = I(1)^b$ for an integer b.

If a parity-check matrix H_C of C has the following form:

$$H_C = \begin{bmatrix} I(c_{0,0}) & I(c_{0,1}) & \cdots & I(c_{0,L-1}) \\ I(c_{1,0}) & I(c_{1,1}) & \cdots & I(c_{1,L-1}) \\ \vdots & \vdots & \ddots & \vdots \\ I(c_{J-1,0}) & I(c_{J-1,1}) & \cdots & I(c_{J-1,L-1}) \end{bmatrix},$$

then C is called a **quasi-cyclic (QC) LDPC code**, where $c_{i,j} \in [P_\infty] := \{0, 1, \ldots, P-1\} \cup \{\infty\}$ and P is the size of a circulant matrix $I(1)$.

Let us denote a matrix which consists of the indices of H_C by $\mathcal{H}_C$, in other words,

$$\mathcal{H}_C = \begin{bmatrix} c_{0,0} & c_{0,1} & \cdots & c_{0,L-1} \\ c_{1,0} & c_{1,1} & \cdots & c_{1,L-1} \\ \vdots & \vdots & \ddots & \vdots \\ c_{J-1,0} & c_{J-1,1} & \cdots & c_{J-1,L-1} \end{bmatrix}.$$

We call $\mathcal{H}_C$ the **model matrix** of H_C.

Imagine that we have a random low-density parity-check matrix. Because of the randomness, we need large memory to store the parity-check matrix. On the other hand, we do not need large memory to keep quasi-cyclic parity-check matrix because the parity-check matrix is reconstructed by its model matrix. Quasi-cyclic LDPC codes are good not only from the viewpoint of memory, but also from the viewpoint of error-correcting performance. In particular, with sum-product decoding, it is expected that quasi-cyclic LDPC codes have good performance for short length codes.[6]

5.4. *Twisted condition for quasi-cyclic LDPC codes*

We give some necessary and sufficient conditions for QC-LDPC codes C and D to satisfy the twisted condition. Let us denote rows of index matrices $\mathcal{H}_C$ and $\mathcal{H}_D$ of quasi-cyclic LDPC matrices H_C and H_D by c_j and d_k, respectively, in other words, $c_j := (c_{j,0}, c_{j,1}, \ldots, c_{j,L-1}), d_k := (d_{k,0}, d_{k,1}, \ldots, d_{k,L-1})$. For Example 5.1, we obtain $c_0 = (1, 2, 4, 2, 4, 1)$ and $d_1 = (6, 5, 3, 5, 3, 6)$.

For an integer sequence $x = (x_0, x_1, \ldots, x_{L-1})$, we call x **multiplicity even** if each entry appears even times in $x_0, x_1, \ldots, x_{L-1}$ except for the symbol ∞. For example, $(0, 1, 1, 0, 3, 3, 3, \infty)$ is multiplicity even, but $(0, 2, 2, 4, 4, 4, 0)$ is not.

Let us define the difference operation "$-$" over $[P_\infty]$ by as follows: for $x, y \in \{0, 1, \ldots, P-1\}$, put

$$x - y := x - y \pmod{P},$$

and

$$x - \infty = \infty - x = \infty - \infty := \infty.$$

Theorem 5.1. *Let C and D be QC-LDPC codes with model matrices $\mathcal{H}_C$ and $\mathcal{H}_D$ respectively such that their size of the circulant matrices, which are components of the parity-check matrices, are the same. The codes C and D satisfy the twisted condition if and only if $c_j - d_k$ is multiplicity even for any row c_j of $\mathcal{H}_C$ and any row d_k of $\mathcal{H}_D$.*

Proof. We divide H_C into J row-blocks $H_{C_1}, H_{C_2}, \ldots, H_{C_{(J-1)}}$:

$$H_{C_j} := (I(c_{j,0}), I(c_{j,1}), \ldots, I(c_{j,L-1})), 0 \leq j < J.$$

Similarly, we divide H_D into K row-blocks. These codes C and D satisfy the twisted condition if and only if $H_C \times H_D^T = 0$ (Proposition 4.1). It is easy to verify that any row of H_C and any row of H_D are orthogonal if and only if any row of H_{C_j} and any row of H_{D_k} are orthogonal for any $0 \leq j < J$ and $0 \leq k < K$.

Denote ath row of H_{C_j} by $\mathcal{C}_a$ and bth row of H_{D_k} by $\mathcal{D}_b$. Thus $\mathcal{C}_a$ and $\mathcal{D}_b$ are binary vectors. Define a binary matrix $X := (x_{a,b})_{0 \leq a,b < P}$, where $x_{a,b} := \langle \mathcal{C}_a, \mathcal{D}_b \rangle = \mathcal{C}_a \times \mathcal{D}_b^{\mathrm{T}}$. Then for $0 \leq j < J$ and $0 \leq k < K$, $X = 0$ if and only if any row of H_{C_j} and any row of H_{D_k} are orthogonal to each other.

By the definition of X, $X = H_{Cj} \times H_{Dk}^{\mathrm{T}}$. And by $I(i)^{\mathrm{T}} = I(-i)$, we have $H_{Cj} \times H_{Dk}^{\mathrm{T}} = \sum_{0 \leq l < L} I(c_{j,l} - d_{k,l})$. Since $I(x)$ is a binary circulant permutation matrix for any integer x, $X = 0$ if and only if $c_j - d_k$ is multiplicity even. $\qquad\square$

5.5. CSS QC-LDPC codes from right-shifted matrices

We call a matrix M a **right-shifted matrix** if M has the following form:

$$
M = \begin{pmatrix}
m_0 & m_1 & \cdots & m_{L-1} \\
m_{L-1} & m_0 & \cdots & m_{L-2} \\
\vdots & & & \vdots \\
m_1 & m_2 & \cdots & m_0
\end{pmatrix},
$$

where $m_i \in [P_\infty]$.

Let A and B be right-shifted matrices over $[P_\infty]$. Let us define model matrices $\mathcal{H}_C$ and $\mathcal{H}_D$ of linear codes C and D respectively by putting: $H_C := [A, B]$ and $H_D := [-B^{\mathrm{T}}, -A^{\mathrm{T}}]$.

Then we have the following propositions:

Proposition 5.1. *The QC-LDPC codes C and D above satisfy the twisted relation.*

Proof. By theorem 5.1, a pair of LDPC codes C and D satisfies the twisted relation if and only if $c_j - d_k$ is a multiplicity even vector for any pair of a row c_j of $\mathcal{H}_C$ and a row d_k of $\mathcal{H}_D$.

Denote the first rows of A and B by $(a_0, a_1, \ldots, a_{L-1})$ and $(b_0, b_1, \ldots, b_{L-1})$, respectively. Then we can rewrite $c_j - d_k = (a_{1-j} + b_{k-1}, a_{2-j} + b_{k-2}, \ldots, a_{L-j} + b_{k-L}, b_{1-j} + a_{k-1}, b_{2-j} + a_{k-2}, \ldots, b_{L-j} + a_{k-L})$. Then the entries of the left-half of $c_j - d_k$ are

$$\{a_0 + b_{k-j}, a_1 + b_{k-j-1}, \ldots, a_{L-1} + b_{k-j-L+1}\},$$

and the entries of the right-half of $c_k - d_j$ are

$$\{a_0 + b_{k-j}, a_1 + b_{k-j-1}, \ldots, a_{L-1} + b_{k-j-L+1}\}.$$

Thus $c_j - d_k$ is a multiplicity even vector. $\qquad\square$

Two linear codes C and D are called **equivalent** if one can be obtained from the other by a permutation of the coordinates.

Proposition 5.2. *The linear codes C and D in Proposition 5.1 are equivalent and $C/D^\perp$ and $D/C^\perp$ are equivalent also.*

Proof. For the parity-check matrices $H_C = (hc_{i,j})$ and $H_D = (hd_{i,j})$, where $hc_{i,j}, hd_{i,j} \in \{0, 1\}$ and $hc_{i,j} = hd_{L-i,2L-j}$ holds. Thus C and D are equivalent.

Let us denote the permutation by σ such that $C = \sigma(D)$. We note that $\sigma(D^\perp) = \sigma(D)^\perp$. □

5.6. *MacKay's code*

The study of LDPC quantum codes was developed by MacKay, who proposed the following construction of a LDPC CSS code:[7]

$$H_C = H_D = \begin{bmatrix} A & A^{\mathrm{T}} \end{bmatrix},$$

where A is a binary right-shifted matrix.

MacKay's code is a special case of the codes derived from Proposition 5.1. In fact, we obtain MacKay's code by putting the size of the circulant matrix $P = 1$ and $B = A^{\mathrm{T}}$.

5.7. *Regular LDPC codes*

An LDPC code is called (ρ, λ)-**regular** if the numbers of 1's in any rows and any columns of a parity-check matrix are constants λ and ρ, respectively. The parameter λ is called the **column weight** and ρ is called the **row weight**. At the beginning of the study of LDPC codes, the construction research has focused on regular LDPC codes. In the design of regular LDPC codes, the following three parameters are regarded important: the column weight λ, the row weight ρ, and **the girth**.

The Tanner graph of a parity-check matrix $\{h_{i,j}\}_{i,j}$ is a bipartite graph with the vertices of *rows* and *columns* of the parity-check matrix and the *edges* between the ith row and the jth column with $h_{i,j} = 1$. The girth is the length of the shortest cycle of the Tanner graph, which is defined below, of a low-density parity-check matrix. It is thought generally that a small girth has bad effects on the performance of sum-product decoding. If the Tanner graph has a short cycle, then the high performance of sum-product decoding is not expected. If the Tanner graph has no cycle, i.e. the graph is a tree, then it is expected that the performance of the sum-product algorithm is the same as that of a maximally likelihood decoding. By the regularity of a parity-check matrix associated with a regular LDPC code, it is unavoidable to contain cycles in the Tanner graph. Thus it is important to attain a big girth in the Tanner graph as one of the research directions (See Refs. 8 and 9 for details).

We would like to leave the following exercise to readers: *compare error-correcting performance of girth-4 LDPC codes with that of girth-6 LDPC codes by computer simulation.*

If $C = D$ and $C^\perp \subset C$, the CSS code obtained by C is called **self-dual containing**.[7,10] For example, the Mackay's code, appearing in the previous subsection, is a self-dual containing code. With quasi-cyclic LDPC conditions, we have the following negative result for self-dual containing.

Proposition 5.3. *Let C be a (λ, ρ) LDPC code with $\lambda \geq 2$. If C is self-dual containing, then the girth of the LDPC code is 4.*

Proof. $C^\perp \subset C$ is equivalent to $H_C \times H_C^{\mathrm{T}}$, where H_C is a parity-check matrix of C. If $\lambda \geq 2$, then there exist rows $h_i = (h_{i,1}, h_{i,2} \ldots,)$ and $h_j = (h_{j,1}, h_{j,2}, \ldots)$ of H_C such that $h_{i,k} = h_{j,k} = 1$ for some k. By assumption $C^\perp \subset C$, we have $h_i \times h_j^{\mathrm{T}} = 0$. On the other hand, $h_i \times h_j^{\mathrm{T}} = \sum_p h_{i,p} h_{j,p} = 1 + \sum_{p \neq k} h_{i,p} h_{j,p}$. Thus there must be an index k' such that $k \neq k'$ and $h_{i,k'} = h_{j,k'} = 1$. $\square$

Remark 5.1. Let C (resp. D) be a classical code associated with MacKay's code and H_C (resp. H_D) a parity-check matrix of the MacKay's code. Then $C^\perp \subset C$ holds. By Proposition 5.3, the girth of the Tanner graph associated with H_C is 4.

On the other hand, the girth of the Tanner graph derived from the construction in Proposition 5.1 is not limited to 4. In fact, we realize the girth-6 codes which satisfy the twisted condition (Theorem 5.3).

In order to construct a code with good error-correcting performance, we should avoid having the size-4 cycle, i.e. the girth to be 4. If this is the case, we denote the following condition by (G):

(G) The girth of the Tanner graph of a parity-check matrix is greater than or equal to 6.

Proposition 5.4 (Ref. 11). *A necessary and sufficient condition to have girth ≥ 6 in the Tanner graph representation of the parity-check matrix of a QC-LDPC code with the index matrix $\{c_{i,j}\}_{i,j}$ is $c_{j_1,l_1} - c_{j_1,l_2} + c_{k_1,l_2} - c_{k_1,l_1} \neq 0$ for any $0 \leq j_1 < j_2 < J, 0 \leq l_1 < l_2 < L$.*

Proof. If there is a sequence $c_{j_1,l_1}, c_{j_1,l_2}, c_{j_2,l_2}$, and c_{j_2,l_1} such that $c_{j_1,l_1} - c_{j_1,l_2} + c_{k_1,l_2} - c_{k_1,l_1} = 0$, then there is a cycle of length 4. In fact, entries

$$H_{1+Pj_1,c_{j_1,l_1}+Pl_1}, H_{1+Pj_1,c_{j_1,l_2}+Pl_2}, H_{1+Pj_2,c_{j_2,l_2}+Pl_2}, H_{1+Pj_2,c_{j_2,l_1}+Pl_1}$$

are all 1's in the parity-check matrix and form a cycle of length 4.
Conversely, if there is a cycle of length 4, which consists of

$$H_{a_1+Pj_1,b_1+Pl_1}, H_{a_2+Pj_1,b_2+Pl_2}, H_{a_3+Pj_2,b_3+Pl_2}, \text{ and } H_{a_4+Pj_2,b_4+Pl_1},$$

then $c_{j_1,l_1} - c_{j_1,l_2} + c_{k_1,l_2} - c_{k_1,l_1} = 0$ holds. □

It is obvious that Proposition 5.4 is equivalent to Proposition 5.5 below. A term "**multiplicity free**" means all the entries of a given vector are different. For example, $(0, 1, 2, 3, 4, 5, 6, 7)$ is multiplicity free, however, $(0, 0, 1, 2, 3, 4, 5)$ is not.

Proposition 5.5. *A necessary and sufficient condition for a QC-LDPC code C to have girth ≥ 6 is $c_{j_1} - c_{j_2}$ be multiplicity free for any $0 \leq j_1 < j_2 < J$.*

Theorem 5.2. *There is no self-dual containing QC-LDPC code which satisfies simultaneously (G) and the twisted condition without ∞ in the index matrix.*

Proof. This statement is already proven in Proposition 5.3. We give another proof in terms of Quasi-Cyclic LDPC code theory.
Assume that C and D are a pair of linear codes which are the ingredients of a self-dual containing CSS code. It implies $D = C$. By Proposition 5.1, $c_j - c_{j'}$ must be multiplicity even for any $0 \leq j, j' < J$. It follows that the difference cannot by multiplicity free. On the other hand, $c_j - c_{j'}$ must be multiplicity free if (G) holds by Proposition 5.5. Therefore (G) and the twisted condition cannot hold simultaneously. □

5.8. *Construction of CSS LDPC codes*

The integers module P, denoted $\mathbb{Z}_P$, is the set of (equivalence classes of) integers $\{0, 1, \ldots, P-1\}$. Addition, subtraction, and multiplication in $\mathbb{Z}_P$ are performed module P.

Proposition 5.6 (Ref. 12). *For any positive integer $P \geq 2$, $\mathbb{Z}_P^* := \{z \in \mathbb{Z}_P | z^{-1}$, which is the inverse element of z i.e. $z(z^{-1}) = (z^{-1})z = 1$, exists$\}$ is an abelian (i.e. commutative) group with multiple operation of $\mathbb{Z}_P$.*

Proof. The commutativity holds obviously. The identity is $1 \in \mathbb{Z}_P^*$. For $z \in \mathbb{Z}_P^*$, the inverse z^{-1} is in $\mathbb{Z}_P^*$ because of $z = (z^{-1})^{-1}$. Finally, for $x, y \in \mathbb{Z}_P^*$, there exists $(xy)^{-1} = x^{-1}y^{-1}$. $\qquad\square$

Proposition 5.7 (Ref. 12). *Let G be a group and H a subgroup of G. Put $[g] := \{hg | h \in H\}$ for $g \in G$. Then we have*
$$[g] \cap [g'] \neq \emptyset \iff [g] = [g'] \iff gg'^{-1} \in H \iff g'g^{-1} \in H.$$

Proof. Let $g, g' \in G$. Assume $[g] \cap [g'] \neq \emptyset$. Then there exist $h, h' \in H$ such that $hg = h'g'$. This implies $g = h^{-1}h'g'$. For any $g_0 \in G$, there exists $h_0 \in H$ such that $g_0 = h_0 g$ by the definition of $[g]$. By $g \in [g']$, we have $g_0 = h_0 g = h_0 h^{-1} h' g' \in [g']$. Thus $[g] \subset [g']$ holds. By a similar argument, $[g] \supset [g']$ holds. Therefore $[g] = [g']$.

If $[g] = [g']$, then there is $h' \in H$ such that $g = h'g'$. This implies
$$g/g' = h' \in H.$$

If $g/g' \in H$, then
$$(g/g')^{-1} = g'/g \in H.$$

If $g'/g \in H$, then there is $h \in H$ such that
$$g'/g = h.$$

Because $g' \in [g']$ and $hg \in [g]$, $g' = hg \in [g'] \cap [g]$. $\qquad\square$

Theorem 5.3. *Let P be an integer > 2 and σ an element of $\mathbb{Z}_P^*$ with the even order $L := \mathrm{ord}(\sigma) \neq \#\mathbb{Z}_P^*$, where $\mathrm{ord}(\sigma) := \min\{m > 0 | \sigma^m = 1\}$. Pick any $\tau \in \mathbb{Z}_P^* \setminus \{1, \sigma, \sigma^2, \dots\}$. Put*

$$c_{j,l} := \begin{cases} \sigma^{-j+l} & 0 \leq l < L/2 \\ \tau\sigma^{-j-1+l} & L/2 \leq l < L, \end{cases}$$

$$d_{k,l} := \begin{cases} -\tau\sigma^{-k-1+l} & 0 \leq l < L/2 \\ -\sigma^{-k+l} & L/2 \leq l < L \end{cases}$$

and put $\mathcal{H}_C = [c_{j,l}]_{0 \leq j < J, 0 \leq l < L}$, $\mathcal{H}_D = [d_{k,l}]_{0 \leq k < K, 0 \leq l < L}$, where $1 \leq J, K \leq L/2$. Then C and D satisfy (G) and the twisted condition.

Example 5.1. Let $P = 7$, ($\mathbb{Z}_P^* = \{1, 2, 3, 4, 5, 6\}$) and choose $\sigma = 2$ and $\tau = 3$. Then $L/2 = \mathrm{ord}(2) = 3$ and we put $J = K = L/2 \ (= 3)$. Then the parity-check matrices are obtained from Theorem 5.3 as:

$$\mathcal{H}_C = \begin{pmatrix} 1\ 2\ 4\ 3\ 6\ 5 \\ 4\ 1\ 2\ 5\ 3\ 6 \\ 2\ 4\ 1\ 6\ 5\ 3 \end{pmatrix}, \mathcal{H}_D = \begin{pmatrix} 4\ 1\ 2\ 6\ 5\ 3 \\ 2\ 4\ 1\ 3\ 6\ 5 \\ 1\ 2\ 4\ 5\ 3\ 6 \end{pmatrix}.$$

Proof. By Proposition 5.1, the twisted condition holds.

Next, we prove that (G) holds for C by Proposition 5.5. Fix any $1 \leq a < b \leq J$. Denote the first $L/2$ entries of $c_a - c_b$ by X, and the remaining $L/2$ entries by Y. Then we have $X = (\sigma^{-a} - \sigma^{-b}) \times (\sigma, \sigma^2, \sigma^3, \ldots, \sigma^{L/2})$ and $Y = (\sigma^{-a} - \sigma^{-b})\tau \times (1, \sigma, \sigma^2, \ldots, \sigma^{L/2-1})$.

Thus the condition that $c_a - c_b$ be multiplicity free is equivalent to $[\sigma^{-a} - \sigma^{-b}] \neq [\tau(\sigma^{-a} - \sigma^{-b})]$ with notation in Proposition 5.7, It is obvious that $[\sigma^{-a} - \sigma^{-b}] \neq [\tau(\sigma^{-a} - \sigma^{-b})]$ if and only if $[1] \neq [\tau]$. By the choice of τ, $[1] \neq [\tau]$ holds. □

We quote the following fundamental facts from group theory and number theory:

Proposition 5.8 (Ref. 12). *If P is a prime, then $\mathbb{Z}_P^*$ is a cyclic group. In other words, there exists $z \in \mathbb{Z}_P^*$ such that $\mathbb{Z}_P^* = \{1, z, z^2, \ldots z^{P-2}\}$.*

Theorem 5.4 (Dirichlet's Theorem[13]). *Let $A_n := b + an$. Then there are infinitely many primes in the series $A_1, A_2, \ldots$ provided $\gcd(a, b) = 1$, where $\gcd(a, b)$ is the greatest common divisor for a and b.*

Theorem 5.5 (Ref. 14). *Let C and D be QC-LDPC codes without ∞ in index matrices.*

(i) For any odd $L > 0$, there is no pair C and D such that both matrices satisfy (G) and the twisted condition simultaneously.

(ii) For any even $L > 0$ and any $1 \leq J, K \leq L/2$, there exists P such that C is a regular (J, L) QC-LDPC code, D is a regular (K, L) QC-LDPC code with the circulant matrix size P, and C and D satisfy both (G) and the twisted condition.

Proof. (i) Let L be an odd integer and C and D QC-LDPC codes. To satisfy the twisted condition for C and D, it is necessary for $c_j - d_k$ be multiplicity even by Proposition 5.1. However, it is impossible since L is odd.

(ii) Let L be an even integer and J and K integers with $1 \leq J, K < L/2$.

In case $L = 2$, put $P = 3$, $\sigma = 1$, and $\tau = 2$. Then Theorem 5.3 provides us with a pair of linear codes which satisfy (G) and the twisted condition.

From here on, we assume $L \geq 4$. Theorem 5.5 holds if there exists a prime $P = 1 + (L/2)n, x > 2$, Since $\mathbb{Z}_P^*$ is a cyclic group of order $(L/2)n$ (Proposition 5.8), there is a generator z. Put $\sigma := z^n$, then we have $\mathrm{ord}(\sigma) = L/2$ and a set $\mathbb{Z}_P^* \setminus \{1, \sigma, \sigma^2, \ldots\}$ is not empty. Thus we can pick τ up from

$\mathbb{Z}_P^* \setminus \{1, \sigma, \sigma^2, \dots\}$. Theorem 5.5 is a consequence of Theorem 5.3. On the other hand, such a prime P exists by Dirichlet's theorem (Theorem 5.4). $\square$

5.9. *Performance of error-correction and minimum distance*

We performed simulations for evaluating error-correcting performance of our codes with sum-product decoding, which is generally used as a decoding method for LDPC codes, and with some parameters (J, L, P, σ, τ), where the symbols J, L, P, σ, τ are defined in Theorem 5.3. We assumed that the channel is binary symmetric and chosen the maximal iteration $l_{\max} = 128$.

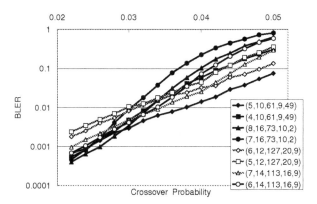

Fig. 3.   Simulation Results: Block error rate (BLER) and Crossover probability of BSC

Figure 3 shows simulation results for our codes C. The list of the parameters (J, L, P, σ, τ) are written bottom right part of the figure. The code length is defined by $L \times P$. The vertical axis shows the block error-rate of the error-correcting performance. Recall that the block error-rate is defined by:

$$\frac{\text{the num. of "error-correction fails"}}{\text{the total num. of simulations}}.$$

The reasons why our simulation results shows the block-error rate of $C/D^\perp$ over the bit-flip binary symmetric channel is the followings:

- The code $D/C^\perp$ is equivalent to $C/D^\perp$ up to the bit-positions. Thus the simulation results of the block-error rate of $D/C^\perp$ over the phase-error symmetric channel is the same as the block-error rate of $C/D^\perp$.
- If we assume that the bit-flip error and the phase-flip error occur independently, the fidelity of the quantum communication over the noisy channel by the quantum CSS code is equal to $\epsilon_B + \epsilon_P - \epsilon_B\epsilon_P$, where ϵ_B and ϵ_P are the block-error rates of $C/D^\perp$ and $D/C^\perp$ over given crossover probabilities respectively.

If $J < L/2$ or $K < L/2$, the minimum distances of classical codes $C/D^\perp$ and $D/C^\perp$ derived from Theorem 5.3 and the related quantum codes cannot exceed L. We know a codeword with the weight L in the parity-check matrix H_C or H_D. Imagine the block error-rate of the related quantum code over the bit-flip crossover rate 0.03 and the phase-flip crossover rate 0.03. If we assume the both errors occur independently, it is expected that approximately 6% of the qubits are flipped by both or one of the errors.

It is known that only half of the minimum distance of the code is correctable by the minimum distance decoding. Remember that the length of our code with ID $(4, 10, 61, 9, 49)$ is 610, the expected number of error is $0.06 \times 610 \simeq 36$, and the minimum distance is at most 10. The minimum distance should be larger than 72. However, we do not have to worry about it since our decoding method is not the minimum distance decoding but the sum-product decoding. See the block error-rate of the code ID $(4, 10, 61, 9, 49)$ in Figure 3. The block error rate of $C/D^\perp$ is less than 0.0048 below the crossover probability 0.03 of the channel. Hence the block error-rate of the quantum CSS code is less than 0.01. Thus it is expected that the block error of the related quantum code occurs only one time per 100 times communications over the bit-flip crossover rate 0.03 and the phase-flip crossover rate 0.03.

6. Postscript

In this review, we discuss quantum error-correcting codes from the viewpoint of a generalization of classical error-correcting codes. In particular, we have focused on the CSS LDPC codes in the second half of this article. Quasi-cyclic LDPC codes are one of the most important classes of classical LDPC codes. In the framework of Quasi-cyclic LDPC codes, it is expected to develop quantum error-correcting coding theory by ring theory, discrete mathematics, combinatorics and others. It is quite a new and interesting subject.

There are several topics on QECC left out in this article. For example, the bound of codes is a meaningful topic for coding theory. **Quantum singleton bound** is the one of famous bounds for quantum error-correcting codes.[3] How many qubits do we need to encode a single qubit such that the quantum error-correcting code overcomes any single quantum error? The answer is $n = 5$.[3] As we have shown in Section 2.1, the answer for classical case is three. Thus the quantum bound is different from the classical bound.

The LDPC codes in this paper are constructed by two classical LDPC codes. There is another generalization of classical LDPC codes to quantum error-correcting codes. That is the class of **stabilizer LDPC codes**. This is a natural generalization of LDPC codes in the view of parity-check measurement of stabilizer code. The interesting problem of stabilizer LDPC codes is a decoding algorithm. At a glance, we find many of cycles of length 4 in the parity-check measurement because of computatibity of observables. As was mentioned in Section 5.7, many cycles of length 4 cause the decoding algorithm failure. Good references of stabilizer codes are Refs. 15,16 and 17.

If an LDPC code is not regular, then the code is called an **irregular** LDPC code. It is known that well-designed irregular classical LDPC codes have better error-correcting performance than regular ones. In fact, it achieves almost the theoretical bound, called the Shannon limit. A similar result is expected for quantum irregular LDPC codes. Indeed, quantum irregular LDPC codes proposed in Refs. 18 show better error-correcting performance than the codes in Section 5.7.

Acknowledgments

The author would like to thank to the staff of Summer School on Mathematical Aspects of Quantum Computing, 27(Mon)-29(Wed) August 2007, at Kinki University, to Dr. Chinen, and to the chief organizer Prof. Nakahara.

This research was partially supported by Grants-in-Aid for Young Scientists (B), 18700017, 2006.

References

1. F. J. MacWilliams and N. J. A. Sloane, *The Theory of Error-Correcting Codes, North-Holland Mathematical Library* (North-Holland, 1977), Chapters 9 and 10.
2. R. G. Gallager, *Low-Density Parity-Check Codes* (Cambridge, MA: MIT, 1963).

3. M. A. Nielsen and I. L. Chuang, *Quantum Computation and Quantum Information*, *Cambridge Series on Information and the Natural Sciences* (Cambridge University Press, 2001).

4. C. E. Shannon, *A Mathematical Theory Of Communication* (Parts 1 and 2), Bell System Technical Journal **27** (1948) 379–423.

5. J. H. van Lint, *Introduction to Coding Theory*, *Graduate Texts in Mathematics* vol. 86 (2nd Edition, Springer Verlag, 1992).

6. LAN-MAN Standards Commitee, http://ieee802.org/

7. D. MacKay, G. Mitchison, and P. McFadden, *Sparse Graph Codes for Quantum Error-Correction*, quant-ph/0304161; IEEE Trans. Inform. Theory **50** (10) (2004) 2315–2330.

8. X.-Y. Hu, E. Eleftheriou, and D. M. Arnold, *Progressive edge-growth Tanner graphs*, in Proceedings of Global Telecommunications Conference 2001 (GLOBECOM '01), vol. 2 (IEEE, 2001) 995–1001.

9. R. M. Tanner, D. Sridhara, and T. Fuja, *A Class of Group-Structured LDPC Codes*, in Proceedings of International Symposium on Communication Theory and Applications (ISCTA, Ambleside, England), (2001).

10. M. Hamada, *Information Rates Achievable with Algebraic Codes on Quantum Discrete Memoryless Channels*, IEEE Trans. Inform. Theory **51** (12) (2005) 4263–4277.

11. M. Fossorier, *Quasi-Cyclic Low-Density Parity-Check Codes From Circulant Permutation Matrices*, IEEE Trans. Inform. Theory **50** (8) (2004) 1788–1793.

12. W. K. Nicholson, *Introduction to Abstract Algebra* (3rd Edition, Wiley, 2006).

13. M. R. Schroeder, *Number Theory in Science and Communication* (4th Edition, Springer, 2005).

14. M. Hagiwara and H. Imai, *Quantum Quasi-Cyclic LDPC Codes*, in Proceedings of 2007 IEEE International Symposium on Information Theory (ISIT) (2007).

15. T. Camara, H. Ollivier, and J.-P. Tillich, *Constructions and performance of classes of quantum LDPC codes*, quant-ph/0502086.

16. J.-P. Tillich, T. Camara, and H. Ollivier, *A class of quantum LDPC code: construction and performances under iterative decoding*, in Proceedings of 2007 IEEE International Symposium on Information Theory (ISIT) (2007).

17. P. Tan and T. J. Li, *On Construction of Two Classes of Efficient Quantum Error-Correction Codes*, in Proceedings of 2007 IEEE International Symposium on Information Theory (ISIT) (2007).

18. M. Hagiwara and H. Imai, *A Simple Construction of Quantum Quasi-Cyclic LDPC Codes*, in Proceedings of 2007 Hawaii and SITA Joint Conference on Information Theory (2007).

CONTROLED TELEPORTATION OF AN ARBITRARY UNKNOWN TWO-QUBIT ENTANGLED STATE

VAHIDEH EBRAHIMI[*], ROBABEH RAHIMI[†], MIKIO NAKAHARA[‡]1

*Interdiciplinary Graduate School of Science and Engineering, Kinki University,
Higashi-Osaka, Osaka 577-8502, Japan*
[1]*Department of Physics, Kinki Univeristy, Higashi-Osaka 577-8502, Japan*
[*]*E-mail: ebrahimi@alice.math.kindai.ac.jp*
[†]*E-mail: rahimi@alice.math.kindai.ac.jp*
[‡]*E-mail: nakahara@math.kindai.ac.jp*

We investigate controlled quantum teleportation of an arbitrary entangled state. It is shown [Man et al., J. Phys. B **40** (2007) 1767] that by using EPR and GHZ states, an unknown N-qubit entangled state can be teleported from a sender to a receiver, under control of a third party. This protocol is studied in more details in our work. The protocol is simulated for a small number of qubits. It is found that a W state, instead of the GHZ state, should be used for a successful teleportation of an entangled state. The difference between the initial state to be teleported and the final state reproduced is studied for different amount of entanglement of the W-like state.

Keywords: Teleportation, Entanglement.

1. Introduction

Quantum information processing is performed by using classical and quantum resources. A classical communication channel is relatively cheap compared to quantum resources, such as entanglement. Therefore, we have to reduce the required quantum resources as much as possible. The quantum resources required are protocol dependent. The majority of the protocols are hard to implement in practice since production of multiqubit entangled states is not easy at present and the difficulty increases considerably with the number of qubits taking part in entanglement.[1]

Recently, Man et al.[2] have shown that an unknown N-qubit entangled state can be teleported from a sender to a receiver, under the control of a third party. By simulating this protocol for a small number of qubits, we have found that a W state is required so that the protocol of the controlled teleportation works fine. For further investigation, we have used W-like entangled states for this protocol. The difference between the initial state for controlled teleportation and the final state reproduced by the receiver is studied for a range of a maximally entangled state to a maximally mixed state. Reasonably, we have found the best agreement between the initial state and the final state is attained if a maximally-entangled W state is employed.

2. Controlled Teleportation of Bipartite Entanglement

Suppose Alice teleports her N-qubit state to Charlie, the receiver, under the M controllers Bob_1, Bob_2, ..., Bob_M. This protocol works for $M < N$ and consumes M GHZ states plus $(N - M)$ EPR pairs.[2] In the controlled teleportation of a two-qubit entangled state according to the original proposal,[2] Alice, Bob, and Charlie share a GHZ state of which the first (the second and the third) qubit is held by Alice (Bob and Charlie) while an EPR pair is shared only between Alice and Charlie.

Analysis of the designed quantum circuit based on the original proposal[2] revealed that Charlie cannot receive entangled two qubits from Alice under the control of Bob, as far as the initial tripartite shared state is a GHZ state. It is due to this fact that tracing out Bob's qubit in the GHZ state leaves the other two-qubit system in a mixed classical state. This result shows that the controlled teleportation of bipartite entanglement cannot be done by using the proposal provided in Ref. 2, while it works correctly by using the other maximally entanglement tripartite state (the W state): Tracing out Bob's qubit in the W state leaves an EPR state shared by Alice and Charlie.

We have found the best agreement between the initial state to be teleported and the final state reproduced by the receiver when the original shared multipartite state is a W state. As the amount of the entanglement of a W-like state decreases, the final state deviates from the expected initial EPR state to be teleported.

References

1. F. G. Deng, C. Y. Li, Y. S. Li, H. Y. Zhou, Y. Wang, *Phys. Rev. A* **72** (2005) 022338.
2. Z. X. Man, Y. J. Xia, N. B. An, *J. Phys. B: At. Mol. Opt. Phys.* **40** (2007) 1767.

NOTES ON THE DÜR–CIRAC CLASSIFICATION

YUKIHIRO OTA

Department of Physics, Kinki university,
Higashi–Osaka, 577–8502, Japan
E-mail: yota@alice.math.kindai.ac.jp

MOTOYUKI YOSHIDA[*,§], ICHIRO OHBA[*,†,‡,¶]

[*]*Department of Physics, Waseda University,*
Tokyo, 169–8555, Japan
[†]*Research Institute for Materials Science and Technology, Waseda University,*
Tokyo, 169–0051, Japan
[‡]*Research Institute for Science and Engineering, Waseda University,*
E-mail: [§]*motoyuki@hep.phys.waseda.ac.jp*
[¶]*ohba@waseda.jp*

One cannot always obtain information about entanglement by the Dür–Cirac (DC) method. The impracticality is attributed to the decrease of entanglement by local operations in the DC method. Even in 2–qubit systems, there exist states whose entangled property the DC method never evaluates. This result is closely related to the discussion of purification protocols. In addition, we discuss its effectiveness in multiqubit systems.

Keywords: Multiparticle Entanglement, Local Operations, Fully Entangled Fraction, Logarithmic Negativity.

1. Introduction

A systematic and effective way to classify multiparticle entanglement in N–qubit systems was proposed by Dür and Cirac.[1] We call it the Dür–Cirac (DC) method. The main idea is explained below. Using a sequence of local operations (LO), one can transform an arbitrary density matrix of an N–qubit system into a state whose entangled property is easily examined. Note that entanglement cannot increase through LO. If the density matrix transformed by LO is entangled, the original one represents an entangled state. However, one cannot always obtain an entangled property by it. We suggested there exists an impracticality in the DC method through an example.[2]

2. Results

First, we explain there exists such impracticality even in 2–qubit systems.[3] One can never determine whether a quantum state is entangled or not by the DC method if the fully entangled fraction[4] is less than or equal to 1/2. This result is closely related to purification protocols. We concentrate on the LO in the recurrence method.[5] The arbitrary density matrix of a 2–qubit system can be transformed into the one which is almost the same as in the DC method. It can be distilled by this method if and only if its fully entangled fraction is greater than 1/2. Next, let us examine its effectiveness in an N–qubit system $(N \geq 3)$. We focus on the following type of density matrices: $\rho_{\mathrm{BD}}^+ = \sum_{j=0}^{2^{N-1}-1} \mu_j^+ |\psi_j^+\rangle\langle\psi_j^+|$, where $\mu_j^+ \geq 0$. The corresponding eigenvectors are given by $|\psi_j^+\rangle = (|0j\rangle + |1\bar{j}\rangle)/\sqrt{2}$ $(\bar{j} = 2^{N-1} - 1 - j)$. Let us consider the density matrix transformed from ρ_{BD} by the LO in the DC method. With respect to an arbitrary bipartition of the system, we can find its logarithmic negativity, which is known as an upper bound of the distillable entanglement,[6] is not vanishing, as far as the maximal eigenvalues of ρ_{BD}^+ are not degenerate. Nevertheless, the strict posotivity of the distillable entanglement for ρ_{BD}^+ has not yet shown.

3. Summary

In a 2–qubit system, one can't always obtain the entangled property of a quantum state by the DC method, if the fully entangled fraction is less than or equal to 1/2. This result is consistent with the consideration on workability of a purification protocol for a 2–qubit. The situation is more complicated in a multiqubit system. We have partially understood its efficiency; one may obtain the entangled property of ρ_{BD}^+.

References

1. W. Dür and J. I. Cirac, *Phys. Rev. A* **61** (2000) 042314.
2. Y. Ota, S. Mikami, M. Yoshida and I. Ohba, Accepted for publication in J. Phys. A; quant-ph/0612158.
3. Y. Ota, M. Yoshida and I. Ohba, arXiv:0704.1375 [quant-ph].
4. G. Alber, T. Beth, M. Horodecki, P. Horodecki, R. Horodecki, M. Rötteler, H. Weinfurter, R. Werner and A. Zeilinger, *Quantum Information: An Introduction to Basic Theoretical Concepts and Experiments* (Springer, Berlin, 2001).
5. C. H. Bennett, D. P. DiVincenzo, J. A. Smolin and W. K. Wootters, *Phys. Rev. A* **54** (1996) 3824.
6. G. Vidal and R. F. Werner, *Phys. Rev. A* **65** (2002) 032314.

BANG-BANG CONTROL OF ENTANGLEMENT IN SPIN-BUS-BOSON MODEL[*]

ROBABEH RAHIMI[1*], AKIRA SAITOH[2†], MIKIO NAKAHARA[1‡]

[1] *Interdiciplinary Graduate School of Science and Engineering, Kinki University,*
Higashi-Osaka, Osaka 577-8502, Japan
[*] E-mail: rahimi@alice.math.kindai.ac.jp
[‡] E-mail: nakahara@math.kindai.ac.jp

[2] *Graduate School of Engineering Science, Osaka University,*
Toyonaka, Osaka 560-8531, Japan
[†] E-mail: saitoh@qc.ee.es.osaka-u.ac.jp

Electron spin bus molecules have been used for implementations of quantum computing, including realizing entanglement. However, systems consisting electron spins are usually unstable. We have studied electron spin bus molecular systems in order to introduce methods with which the produced states i.e. entangled states, stay stable. According to the operational evidences, the problem is numerically studied by introducing a model of decoherence and applying fast multipulses for suppressing decoherence. We found that bang-bang pulses with realistic duty ratios can somehow suppress decoherence.

Keywords: Entanglement, Decoherence, Electron Spin Bus Quantum Computing.

1. Introduction

Molecular systems involving electron spins have been used for implementations of quantum computing, specially for realizing quantum entanglement.[1-3] One of the advantages of this system for realizing a quantum computer is that entanglement between an electron spin and a nuclear spin can be achieved under milder experimental conditions compared to the case of an eg. NMR system where entanglement is being realized between two nuclear spins. However, the physical system involving an electron spin is usually very fragile since an electron spin possesses very short decoherence time. In an attempt to remove this drawback , we have studied the electron spin bus systems.

---

[*] A precise discussion can be found in J. Phys. Soc. Japan **76** (2007) 114007.

2. The Model of Decoherence

In a conventional model boson modes are distributed uniformly over all spin systems in a solid material. In fact, this model ignores the natural dissipation of bosons through a cavity and/or a sample holder. Then under the influence of noise, by using a conventional noise model, oscillations of entanglement rather than a decay of entanglement is found since the total system involving the spins and the bosonic bath is a closed system, which is integrable. We consider another model in which bosons dissipate into an extra bath of bosons of a cavity and/or a sample holder. According to the operational evidences, the dynamic of entanglement for an electron spin bus system is numerically solved by employing the introduced model of decoherence and applying fast multipulses for suppressing decoherence.

3. Results

It is mentioned that the system we deal with is particularly useful for making an entangled state. This is due to the fact that high spin polarization can be transferred from a polarized electron spin to a less polarized nuclear spin. The total state after transmission, would be highly polarized, thus appropriate for generalizing entanglement. Here, we study two cases of one with polarization transfer and else without any polarization transfer pulse.

In a numerical study, we take a set of realistic parameters representing a commercial Q-band machine (magnetic field of 35 GHz for an electron nuclear double resonance, ENDOR, machine), that is currently available for our experiment. In case no polarization transfer is used, a high negativity is achieved when the temperature is less than 10 mK; if, in contrast, the ideal polarization transfer is used, a high negativity is achieved when the temperature is even at around 10^3 mK.

Totally we conclude that decoherence matter is rather more challenging, while entanglement is comparatively easy to be established for an electron spin bus molecule system for quantum computing. When number of qubits is small, we still find some regions of parameters where the quantum state can be stable, if careful tuning of bang-bang pulses are made.

References

1. M. Mehring, J. Mende, W. Scherer, *Phys. Rev. Lett.* **90** (2003) 153001.
2. R. Rahimi: PhD Thesis, Osaka Univ., 2006, quant-ph/0609063.
3. W. G. Morley, J. van Tol, A. Ardavan, K. Porfyrakis, J. Zhang, G. A. D. Briggs, quant-ph/0611276.

NUMERICAL COMPUTATION OF TIME-DEPENDENT MULTIPARTITE NONCLASSICAL CORRELATION

AKIRA SAITOH, ROBABEH RAHIMI[†],

MIKIO NAKAHARA[†], MASAHIRO KITAGAWA

Department of Systems Innovation, Graduate School of Engineering Science, Osaka University, 1-3 Machikaneyama, Toyonaka, Osaka 560-8531, Japan
[†] Interdisciplinary Graduate School of Science and Engineering, Kinki University, 3-4-1 Kowakae, Higashi Osaka, Osaka 577-8502, Japan

Non-classical correlation of a multipartite quantum system is an indispensable constituent of quantum information processing (QIP).[1] There is a commonly recognized paradigm[2] in which a multipartite quantum system described by a density matrix having no product eigenbasis is considered to possess nonclassical correlation. The set of such density matrices includes the set of entangled states as shown in Fig. 1(a). Effective QIP requires quantumness to be stable against noise.

There is a long-term study of suppressing decoherence in quantumness. Here, we focus on the bang-bang control scheme[3] together with a sort of spin-boson linear coupling models. Consider a spin system S (consisting of

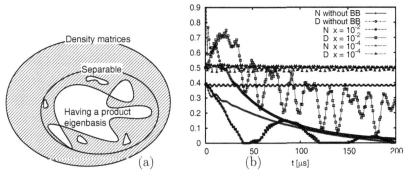

Fig. 1. (a) Set of density matrices having no product eigenbasis (shaded region). Its complementary set is non-convex. (b) Time evolutions of D and negativity (N) with/without bang-bang control (x is the duty ratio tried in the numerical simulation).

n spins $1/2$) coupled with a on-resonance bosonic system B replaced with a thermal one with dissipation probability p for a time interval τ. The system Hamiltonian is set to $H = H_S + H_B + H_c$ with $H_S = \sum_{j=1}^{n} f_j S_j + \sum_{j,k,j<k} A_{jk} S_j S_k$ (spin system Hamiltonian), $H_B = \sum_{j=1}^{n} f_j a_j^\dagger a_j$ (Hamiltonian of bosonic modes), and $H_c = \sum_{j=1}^{n} c_j S_j (a_j^\dagger + a_j)$. Here, f_j is the precession frequency of the jth spin ($f_j \neq f_k$ for $j \neq k$), A_{jk} is a spin-spin coupling constant, c_j is a spin-boson on-resonance coupling constant. We consider a bang-bang pulse irradiation with pulse width L and duty ratio x [i.e., each X_π pulse is irradiated with the duration xL followed by the absence time interval $(1 - x)L$]. The bang-bang pulses for individual spins are synchronized.

We numerically evaluate the effect of bang-bang pulses on the time evolution of a nonclassical-correlation measure defined as follows. Consider a density matrix $\rho^{[1,\cdots,m]}$ of an m-partite system. Consider local complete orthonormal bases $\{|e_j^{[1]}\rangle\}_j, \ldots, \{|e_z^{[m]}\rangle\}_z$. Then, a measure of nonclassical correlation of our interest is defined by

$$D(\rho^{[1,\cdots,m]}) = \min_{\text{local bases}} \left(-\sum_{j,\ldots,z} p_{j,\ldots,z} \log_2 p_{j,\ldots,z} \right) - S_{\text{vN}}(\rho^{[1,\cdots,m]})$$

with $p_{j,\ldots,z} = \langle e_j^{[1]} | \langle e_k^{[2]} | \cdots \langle e_z^{[m]} | \rho^{[1,\cdots,m]} | e_j^{[1]} \rangle | e_k^{[2]} \rangle \cdots | e_z^{[m]} \rangle$. The value of $D(\rho^{[1,\cdots,m]})$ is zero if $\rho^{[1,\cdots,m]}$ has a (fully) product eigenbasis. In addition, $D(\rho^{[1,\cdots,m]})$ is invariant under local unitary operations.

One of simulation results is illustrated in Fig. 1(b). For this simulation, we set the following parameter values: $n = 2$, $f_1 = 3.40 \times 10^{10}$[Hz], $f_2 = 4.87 \times 10^7$[Hz], $A_{12} = 1.00 \times 10^7$[Hz], temperature 1mK, $\tau = 1 \times 10^{-6}$[s], $L = 1.0 \times 10^{-7}$[s], and $p = 0.1$. We apply an entangling operation, an Hadamard gate followed by a CNOT gate, to the spins originally in the thermal state to make the initial state for $t = 0$. Time evolutions of D and negativity (one of entanglement measures) with/without bang-bang control are shown in the figure. Suppression of decoherence is significant for the duty ratio $x = 1.0 \times 10^{-4}$.

References

1. J. Gruska, *Quantum Computing* (McGraw-Hill, London, 1999); M. A. Nielsen and I. L. Chuang, *Quantum Computation and Quantum Information* (Cambridge University Press, Cambridge, 2000).
2. J. Oppenheim, M. Horodecki, P. Horodecki, and R. Horodecki, *Phys. Rev. Lett.* **89** (2002) 180402.
3. M. Ban, *J. Mod. Opt.* **45** (1998) 2315; L. Viola and S. Lloyd, *Phys. Rev. A* **58** (1998) 2733.

ON CLASSICAL NO-CLONING THEOREM UNDER LIOUVILLE DYNAMICS AND DISTANCES

TAKUYA YAMANO*, OSAMU IGUCHI

Department of Physics, Ochanomizu University,
2-1-1 Otsuka, Bunkyo-ku, Tokyo 112-8610, Japan
* E-mail: yamano@skycosmos.phys.ocha.ac.jp

We have reported that the classical counterpart of the quantum no-cloning, which has been shown for the Kullback-Leibler distance under Liouville dynamics, can be proven also by non Csiszár f-divergence type distance.

Keywords: No-Cloning, Liouville Dynamics, Distance Measure.

No-cloning theorem[1] tells that no operation $\mathbb{O}$ which allows devices to copy an arbitrary unknown states $\mid \Psi \rangle_s$ to a target states $\mid 0 \rangle_t$ exists,

$$\mid m \rangle \otimes \mid \Psi \rangle_s \otimes \mid 0 \rangle_t \xrightarrow{\mathbb{O}} \mid m_\Psi \rangle \otimes \mid \Psi \rangle_s \otimes \mid \Psi \rangle_s \qquad (1)$$

where the symbols s, m, and t denote source, machine and target, respectively. It has been proven that this theorem has a classical analogue,[2,3] where the crucial logic of the proof is to show a contradiction. That is, the Kullback-Leibler (KL) distance (a special case of the Csiszár f-divergence[4]) is conserved under the Liouvile dynamics (LD) $\partial_t \mathcal{P} = -\nabla \cdot (\vec{v}\mathcal{P})$, where $\vec{v}$ is the velocity associated with generalized coordinates in phase space, on the contrary, the KL distances before and after the process were shown to be not equivalent. To what extent is this analogue robust? Is this peculiar to the KL distance? The situation is not so obvious as the quantum case, where the von Neumann relative entropy $S(\sigma \mid \rho) = Tr(\sigma \log \sigma - \sigma \log \rho)$ between two states σ and ρ is conserved under unitary evolution $\rho(t) = U(t)\rho(0)U^\dagger(t)$ etc. To answer this question, we have investigated whether or not the contradiction exists for the different class (non Csiszár f divergence type) of distance.

The distance is defined as $C_\alpha = -\log[\int d\vec{x} \mathcal{P}_1^\alpha \mathcal{P}_2^{1-\alpha}]$, where $0 \le \alpha \le 1$ and $\mathcal{P}_{1,2}$ are two different statistical ensemble distributions of the system with phase space coordinate $\vec{x} = (\vec{x}_m, \vec{x}_s, \vec{x}_t)$. $C_\alpha(\mathcal{P}_1, \mathcal{P}_2) \ge 0$ holds with equality if and only if $\mathcal{P}_1 = \mathcal{P}_2$ when $\alpha \neq 0$ and $\alpha \neq 1$. We can obtain that the derivative of C_α vanishes when $\mathcal{P}_{1,2}$ obey the LD, i.e., $dC_\alpha(\mathcal{P}_1, \mathcal{P}_2)/dt = 0$.[5] This means that we can expect the conservation of the distance as $C_\alpha(\mathcal{P}_1, \mathcal{P}_2) = C_\alpha(\mathcal{Q}_1, \mathcal{Q}_2)$ for two corresponding different final (after copying) distributions $\mathcal{Q}_{1,2}$. This expectation, however, is found to be unachievable. The factorization of the initial distributions $\mathcal{P}_{1,2} = \mathcal{P}_m(\vec{x}_m)\mathcal{P}_s^{1,2}(\vec{x}_s)\mathcal{P}_t(\vec{x}_t)$ are assumed in the copying process. The marginalization for the final distribution with respect to the coordinate m is a premise i.e., $\int d\vec{x}_m \mathcal{Q}_{1,2} = \mathcal{P}_s^{1,2}\mathcal{P}_t^{1,2}$. The successful cloning produces $\mathcal{Q}_{1,2} = \mathcal{Q}_m \mathcal{P}_s^{1,2}\mathcal{P}_s^{1,2}$ as the final distribution, which corresponds to Eq.(1). Noting that for the two factorized distributions the total distance is a sum of the distances of component systems, we obtain $C_\alpha(\mathcal{P}_1, \mathcal{P}_2) = C_\alpha(\mathcal{P}_s^1, \mathcal{P}_s^2)$ by $C_\alpha(\mathcal{P}_m, \mathcal{P}_m) = 0$ and $C_\alpha(\mathcal{P}_t, \mathcal{P}_t) = 0$. This means that the distance between $\mathcal{P}_1$ and $\mathcal{P}_2$ is attributed to the distance between the probabilities of the sources. With the help of the Hölder inequality, the distance after the copying is then evaluated as

$$C_\alpha(\mathcal{Q}_1, \mathcal{Q}_2) \geq -\log\left(\int d\vec{x}_{s,t}\left[\int d\vec{x}_m \mathcal{Q}_1\right]^\alpha \left[\int d\vec{x}_m \mathcal{Q}_1\right]^{1-\alpha}\right)$$

$$= -2\log\left(\int d\vec{x}_s (\mathcal{P}_s^1)^\alpha (\mathcal{P}_s^2)^{1-\alpha}\right) = 2C_\alpha(\mathcal{P}_1, \mathcal{P}_2). \tag{2}$$

This clearly shows the discrepancy. In summary, the assertion of the classical counterpart of the no-cloning theorem is achieved based on a non-Csiszar's f-divergence distance under the Liouville dynamics. The copying machine cannot produce a statistical ensemble clone under this dynamics. The more details of the present consideration will be provided elsewhere.[5]

Acknowledgments

The authors thank Dr. T. Ootsuka for inviting to participate in the Summer school at Kinki University, Japan.

References

1. W. K. Wootters, W. H. Zurek, *Nature* **299** (1982) 802.
2. A. R. Plastino, A. Daffertshofer, *Phys. Rev. Lett.* **93** (2004) 138701.
3. A. Daffertshofer, A. R. Plastino, A. Plastino, *Phys. Rev. Lett.* **88** (2002) 210601.
4. I. Csiszár, *Studia Math. Hungarica* **2** (1967) 299.
5. T. Yamano, O. Iguchi, Classical no-cloning under Liouville dynamics by non-Csiszár f-divergence, in preparation.